PRAISE FOR
THE HUMAN AGENTIC AI EDGE

"Rather than chasing AI hype, Andreas Welsch focuses on what actually makes a difference for enterprise teams. What we need most on AI projects right now are these kinds of real-world lessons. This book proves that a winning AI strategy is about combining human expertise with Agentic AI, rather than overreaching with tech. If enterprise leaders want their organizations to be AI-ready, they need a thoughtful, pragmatic roadmap, and they get that here."

*— **Jon Reed**, CEO & Co-Founder, diginomica*

"Andreas has written the playbook for leaders navigating AI adoption and high-quality results. By grounding autonomy in ownership, governance, and outcomes, the book moves beyond hype toward sustainable impact. It gives executives a clear, actionable model for turning Agentic AI into a disciplined capability rather than an experiment."

*— **Dr. Marek Kowalkiewicz**, Professor & Chair in Digital Economy,*
Queensland University of Technology

"The best technology won't reveal its true potential unless team members are empowered to use it and are aware of how to use it. Agentic AI promises to fundamentally change how businesses operate, and this book prepares you to shape your organization to be ready to shape this future."

*— **Dr. Sean Kask**, Chief AI Strategy Officer, SAP*

"Whether you are just getting started with Agentic AI or already have an initiative underway, you will find a wealth of information that you can directly apply in your work, no matter the size of your organization."
— *Ivo Strohhammer*, *AI Lead, Siemens Smart Infrastructure and CEO, AI Cluster Zug*

"This book is highly relevant for leaders at every level. Andreas Welsch makes clear that Agentic AI success depends on leaders who can combine human judgment, accountability, and clear standards. These skills can no longer be taken for granted—they must be intentionally cultivated."
— *Yasemin Bozbey*, *Head of Learning & Development, EGGER Group*

"This book speaks directly to the questions business and IT leaders face as Agentic AI enters core business processes. Andreas clearly articulates how architecture, governance, and human accountability must work together to ensure scalable, trustworthy AI-enabled systems."
— *Paul Kurchina*, *Enterprise Architect Community Leader*

"Performance gains from Agentic AI depend more on human judgment than on technology and automation. Rather than chasing automation for its own sake, Andreas focuses on building AI-ready teams with the right skills, standards, and leadership oversight. This book offers a practical blueprint for leaders who want results with accountability."
— *Claudia Saleh*, *Senior AI Manager, Media & Entertainment Industry*

"Andreas perfectly captures how leaders and their teams can move from using AI as a tool to adapting work with Agentic AI while providing actionable advice on how to keep humans at the center."

— **Nico Bitzer**, *CEO & Co-Founder, Bots & People*

"Rather than treating AI as a skills checklist, Andreas frames it as a leadership and capability challenge. The emphasis on responsible and accountable AI use makes it essential reading for L&D leaders."

— **June Huynh**, *AI Learning & Development Lead, Financial Services Industry*

"This book offers leaders and professionals the concrete steps and examples needed to confidently embrace the new ways of working at a time when C-suite executives demand tangible value from Agentic AI."

— **Tobias Zwingmann**, *Managing Director, RAPYD.AI*

"Andreas shows how to combine Agentic AI with human expertise to improve consistency, accuracy, and outcomes across complex processes. The emphasis on workforce readiness and clear responsibility reflects the business reality where AI must perform reliably every day."

— **Ian Barkin**, *Co-Founder & CEO, magentIQ*

"By emphasizing human-in-the-loop decision-making, accountability, and clear quality standards, Andreas Welsch shows how organizations can achieve consistent results without compromising trust."

— **Sadie St Lawrence**, *CEO, Human Machine Collaboration Institute*

THE

HUMAN

AGENTIC

AI

EDGE

THE

HUMAN

AGENTIC

AI

EDGE

SHAPE THE NEXT GENERATION OF *AI-READY TEAMS*

ANDREAS WELSCH

THE HUMAN AGENTIC AI EDGE
Shape the Next Generation of AI-Ready Teams

Andreas Welsch

For more information: andreas@intelligence-briefing.com

ISBN (paperback): 979-8-9936153-0-1
ISBN (ebook): 979-8-9936153-1-8
ISBN (audiobook): 979-8-9936153-2-5

FOREWORD

O rganizations have been built on a stable assumption: work is done by people, supported by tools. AI is breaking that assumption now. The boundary between human and machine work is beginning to blur as software gains autonomy and acts with limited human supervision. As a result, judgment, responsibility, and accountability inside organizations are changing as well. This shift introduces both leverage and fragility as decisions are influenced earlier, actions happen faster, and habits form before leaders recognize a new operating pattern. Leaders share a responsibility to their employees, organizations, and stakeholders to shape how AI and AI agents are introduced, governed, and incorporated into everyday work. Most leaders find themselves in unfamiliar territory, requiring them to make decisions about autonomy, oversight, and accountability without established playbooks or precedents.

Just a few years ago, AI was the private domain of individual businesses with access to vast budgets, pools of data, and highly specialized experts. Since then, AI has quickly become ubiquitous in many parts of our lives. It's available at our fingertips on a mobile device, as an icon in your office productivity tool, or to quickly summarize online search results. Over this timeframe, this technology has also become far more capable for businesses, making it easier and quicker to experiment with and explore.

Agentic AI moves beyond predicting product demand, recommending the best-matching products, or simply automating repeatable steps in business processes. As in earlier innovation cycles, many leaders still believe that focusing on the latest tools is sufficient to create sustainable business value. In fact, they miss the most impactful lever: how their organization and operating model need to evolve as team members use the tools available to them. We have seen this pattern before with other technologies. Companies large and small have gone from "Internet-first" to "Big-Data-first" to "Digital-first." For most, success has been muted. The reason for that is underestimating the complexity of change.

Productivity suffers rather than flourishes when CEOs push "AI-first" narratives, threaten adopt-or-exit measures, and leave employees to explore AI tools on their own. Instead, leaders need to establish clear standards, accountability, and readiness as operating principles when Agentic AI takes on a larger share of knowledge work. Employees quickly develop habits for using the technology. Without standards for responsible use, organizations end up with low-quality results, AI workslop. Without accountability, shortcuts become the norm, and your team's or company's reputation degrades. Without AI readiness, everyone just muddles through or conceals their use of AI out of fear, exposing your company to security and data privacy risks and additional technical debt of an unmanaged shadow AI workforce.

FOREWORD

As a management consultant, I have advised clients during several innovation waves. This time around, the pace is much faster, and the technology is much more capable. In the mid-2010s, we automated pipeline safety inspections with a mix of automation and sophisticated AI models. AI agents can now do this work, recommend actions, and even take them based on the risk level, in a fraction of the time and at a fraction of the operational cost. But this development is not limited to a single business function. At this critical juncture, business and technology leaders need to actively shape how their teams use Agentic AI and make it integral to their operating model, while augmenting their people.

This is what Andreas so succinctly describes as the HUMAN Agentic AI Edge. Whether you are navigating the organizational implications of Agentic AI, redefining the human value of work, or developing an AI charter with your team, you will find frameworks and examples in this book that you can immediately apply.

You may choose to read this book sequentially or begin with the chapters that are most relevant to the challenges you face today. Either way, it serves as a leadership guide for building AI-ready organizations that combine technological capability with human judgment, accountability, and trust. I hope you will enjoy reading *The HUMAN Agentic AI Edge* as much as I have. Let this book be the spark for you to build the AI-ready teams the future demands!

— *Chris Johannessen,* Editor,
Journal of AI, Robotics and Workplace Automation

ABOUT THE AUTHOR

Andreas Welsch is an internationally recognized AI leader, advisor, speaker, and trainer known for helping the world's leading enterprises turn technology hype into business outcomes.

He is the Founder and Chief Human Agentic AI Officer at Intelligence Briefing, a best-selling author, and a leading voice on AI strategy, adoption, and organizational readiness, equipping teams through hands-on programs that develop Certified AI Leaders. Previously, in senior leadership roles at SAP, he has advised Fortune 500 leaders on realizing business value from AI and accelerated the integration of AI across enterprise applications.

Outside of work, he contributes to academia as an Adjunct Professor at West Chester University of Pennsylvania and serves on the Editorial Board of the Journal of AI, Robotics and Workplace Automation.

Andreas is a sought-after keynote speaker and the creator of several LinkedIn Learning leadership courses on Agentic AI. He has been named a LinkedIn Top Voice, a Thinkers360 Top 10 Thought Leader in Agentic AI, and a SwissCognitive Top 50 Global AI Ambassador. Explore his *AI Leadership Handbook*, podcast, *What's the BUZZ?*, and subscribe to *The AI MEMO*. To engage Andreas for keynotes, advisory, or upskilling, visit www.intelligence-briefing.com.

CONTENTS

CONTENTS

WHO SHOULD READ THIS BOOK

I f you are a business or technology leader under growing pressure to adopt Agentic AI while protecting quality, accountability, and trust, this book is for you. If you lead teams or influence how work gets done in finance, procurement, marketing, sales, service, supply chain, human resources, or IT, you will find practical guidance you can apply immediately.

Whether you are a manager, director, C-level executive, or founder, you are likely expected to deliver measurable productivity gains with AI. At the same time, you notice unintended consequences: more drafts to review, generic or inconsistent output, blurred ownership, or uncertainty about what "good" looks like when AI contributes to the work. This book helps you navigate that tension by showing you how to lead AI adoption intentionally.

You will learn how to decide which tasks to delegate to AI, where human judgment must remain central, and how to set clear standards so productivity gains do not come at the expense of trust or reputation. It is a leadership playbook for shaping AI-ready teams that know when to rely on AI, when to intervene, and how to preserve the HUMAN Edge as work evolves. However, it is not a guide to building AI agents or a deep dive into individual technical topics, tools, or frameworks.

INTRODUCTION

A I is no longer a technology storyline, but an operating reality. Right now, most organizations are scaling capability faster than they are scaling judgment. Every week brings new announcements, ranging from larger models to new agent features, and new promises of autonomy. But inside companies, something quieter is happening. Teams are producing more and trusting less. Leaders are drowning in AI-generated drafts, summaries, and insights that appear finished but require review, correction, and risk assessment before anyone can act.

That is why this book is built on three operating principles that we will return to throughout. I call them the *HUMAN Agentic AI Edge Triad*: *standards* that define what "good" means, *accountability* for outcomes when AI contributes, and *readiness*, including the skills, governance, and workflow design that scale quality. When these three lag accelerating AI capabilities, productivity gains become rework, reputational risk, and a growing trust gap.

Instead, leaders need to shape AI-ready teams with a HUMAN Agentic AI Edge through clear leadership, hybrid team design, and accountability mechanisms that scale Agentic AI without scaling mediocrity across the organization and its stakeholders.

AI labs are outpacing one another in delivering new product features, often within days of each other. Established software vendors have also been accelerating their product roadmaps to keep pace, evolving business software from users clicking through workflows to systems that orchestrate work on a user's behalf.

As software companies push the technology into the market, a new shift is underway. When AI becomes the interface across tools and processes, operational trust becomes the hard part. Building that trust requires defining what agents are allowed to do, what "good" results look like, how to handle exceptions, and who is accountable when automation touches customer-facing decisions.

Vendors are pushing innovations faster into the market than their customers can adopt them. Large organizations have the resources to explore and adopt these capabilities, while mid-sized companies take a deliberate or even a wait-and-see approach. Those who delay exploration and adoption risk being left behind so far that they can no longer catch up. Emerging AI-native companies exacerbate this problem because of their agility and lack of legacy systems, processes, and customer expectations.

Boards pressed CEOs during the Generative AI hype to articulate their company's AI strategy. In a rush to explore the technology and satisfy the board's requirement, organizations have quickly deployed AI assistants, tools, and platforms. Since then, software vendors have delivered additional innovations, such as AI agents that act based on a user's goal rather than rigid instructions.

Now, boards are asking harder questions about the quantifiable business impact and benefits that these investments enable. As a result, leaders frequently find themselves caught between the hype around the technology's capabilities and the reality of their organization's constraints, which determine how quickly those capabilities translate into real business outcomes.

Despite tech CEOs pointing to AI-driven gains as a reason for rounds of layoffs since the pandemic, the picture is more nuanced. AI is often used as a welcome excuse to reduce payroll costs or to fund capital-intensive growth through upfront investments in hardware and data centers. Across industries, leaders have slowed or stopped hiring for entry-level roles and for software engineers, in a nod to expected AI-driven efficiencies.

According to market analysts Forrester in 2025[1], more than half of the companies that have laid off staff in anticipation of efficiency gains will engage in quiet rehiring to backfill roles where AI (or its implementation) has not delivered the expected impact. Although AI's technological advancements are in the headlines, leaders and organizations need to pay special attention to the *human* aspects to ensure AI realizes its full value within the organization.

AI use in businesses continues to increase, but so does the number of low-quality, AI-generated messages and content filling

---

[1] Forrester, 2025, "Predictions 2026: The Future Of Work," October 21, 2025, https://www.forrester.com/report/predictions-2026-the-future-of-work/ RES185020.

email inboxes and requiring review and editing before they can be shared with business stakeholders.

Top management expects employees to become more productive with the AI tools the company has bought. Some executives even set productivity targets as high as 20–30%. That pressure can turn AI into a crutch. *"Done is better than perfect"* becomes the default, even as everyone still expects quality. Leaders need to set clear standards, specify when AI is appropriate, and provide teams with hands-on training. Otherwise, generic, bland, and incorrect information will diminish the team's reputation and results, or even reflect on the company, as the first cases illustrate. To prevent it, apply the Labor-Shift Check: If AI reduces your effort but increases someone else's burden to verify the result, you transfer labor rather than increase productivity.

In conversations, leaders and industry experts alike quickly realize that the process of organizational AI adoption is not new. Companies have been here before in earlier innovation cycles. So, what is truly different this time? Research published just before the recent Generative AI boom provides some helpful hints[2].

Organizations and the people within them frequently forget key learnings from one technology wave to the next as leaders move on at the end or halfway through, and systematic post-mortem processes that close the learning loop are ignored. When the next

---

[2] Saghafian, Mina, et al., 2021, "Stagewise Overview of Issues Influencing Organizational Technology Adoption and Use," March 16, 2021, https://www.frontiersin.org/journals/psychology/articles/10.3389/fpsyg.2021.6301 45/full.

innovation cycle begins, change fatigue sets in across the organization, leaders juggle competing priorities, and short-term performance pressures lead decision-makers to repeat past mistakes in pursuit of perceived speed or convenience.

In this process, leaders tend to rely on their intuition or prior success models, even when new circumstances make those models ineffective or obsolete. In essence, individuals do not learn from past mistakes because those mistakes (and learnings) were not their own. That is why learning agility and adaptability are among the most essential skills when change is constant.

In 2025, the World Economic Forum projected that 39% of workers' skills would change within the next five years, largely due to AI[3]. Yet companies dedicate just 7% of their AI budget to upskilling and hands-on training, compared to 93% to technology, as Deloitte found in the same year[4]. But 70% of AI's success depends on people-related aspects such as talent management, change management, and processes, as Boston Consulting Group reported in 2024[5]. Too often, the operational responsibility for shaping AI-ready teams falls to their leaders, who must manage this transition as yet another task on top of their day-to-day

---

[3] World Economic Forum, 2025, "Future of Jobs Report 2025," January 07, 2025, https://www.weforum.org/publications/the-future-of-jobs-report-2025.

[4] Fortune, 2025, "Deloitte's CTO on a stunning AI transformation stat: Companies are spending 93% on tech and only 7% on people," December 15, 2025, https://fortune.com/2025/12/15/deloitte-cto-bill-briggs-what-really-scares-ceos-about-ai-human-resources.

[5] BCG, 2024, "The Leader's Guide to Transforming with AI," December 12, 2024, https://www.bcg.com/featured-insights/the-leaders-guide-to-transforming-with-ai.

responsibilities. Without concrete guidance and examples at this critical moment, only a few succeed.

Organizations are racing to deploy Agentic AI that completes goals on a user's behalf. Yet few are ready for the risks that emerge when employees use AI without structure, standards, or oversight. Instead of driving efficiency, teams are drowning in AI-generated noise disguised as information. Generic content, fabricated facts, and poor decisions are the signs of AI workslop when employees rely on AI without proper guidance, producing low-quality output that slows everyone down and tarnishes your company's reputation. But your customers and stakeholders expect more from you than AI slop at scale. This book offers you a practical blueprint for building accountable AI-ready teams that consistently produce high-quality results. In these chapters, you will discover how to:

- Develop essential AI skills across your workforce
- Combine human expertise with Agentic AI scale
- Establish clear accountability for AI-augmented work
- Normalize responsible, accurate, and trusted AI use
- Scale beyond personal productivity without creating AI slop

You will learn everything you need to know to shape the next generation of AI-ready teams that deliver high-quality results with high accountability and a HUMAN Edge.

Since publishing the *AI Leadership Handbook* in September 2024, I have interviewed more than 50 AI leaders on my podcast, *What's the BUZZ?*, published dozens of thought pieces, and discussed best practices with AI leaders at industry conferences. The range of topics has been similarly expansive, covering the *evolution of leadership, the future of work, upskilling, pilot programs, and Agentic AI*, as well as establishing *governance and security for enterprise AI*. These topics are also central in my online courses on LinkedIn Learning and the *Certified AI Leader*™ flagship courses on www.intelligence-briefing.com.

This book is structured into three parts that cover the HUMAN Edge in using Agentic AI, building AI-ready organizations, and scaling Human-Agentic AI work across teams.

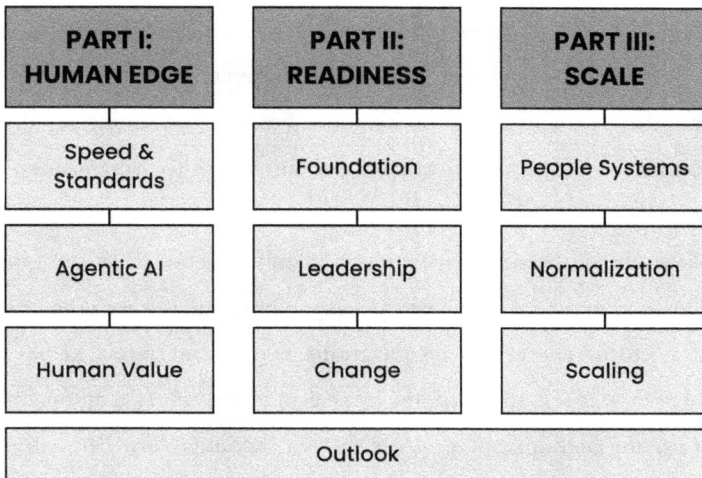

PART I: HUMAN EDGE	PART II: READINESS	PART III: SCALE
Speed & Standards	Foundation	People Systems
Agentic AI	Leadership	Normalization
Human Value	Change	Scaling
Outlook		

Shaping the Next Generation of AI-Ready Teams

In Part I, we will lay the foundation for creating a HUMAN Agentic AI Edge, spanning the background and effects of accelerating AI adoption, the introduction of AI agents, and the impact on roles and the definition of work.

In Part II, we will explore how to shape your AI-ready organization by focusing on the people working within it. After redefining expertise and knowledge, actionable steps for designing and leading hybrid teams of humans and AI agents enable you to put ideas into practice.

In Part III, we will discuss approaches for scaling Human-AI collaboration across your organization. From building on established HR concepts to manage the Agentic AI lifecycle to normalizing AI and scaling its use across your teams, these tips and frameworks will prepare you to encourage and empower your teams to use AI without compromising accountability or quality.

The book closes with an outlook on several Agentic AI topics and their potential implications for business, spanning agentic commerce, browsers, and emerging security risks for organizations.

For CEOs, managers, and entrepreneurs, this book offers a blueprint for the skills, structures, and culture required to combine human judgment with Agentic AI capabilities and to reach the level of performance many organizations expect but rarely achieve. Prepare yourself to shape the next generation of AI-ready teams delivering high-quality results with high accountability. But before you can guide your team into the future, you need to know how we got to the edge of HUMAN Agentic AI and where to go from here.

PART I

THE HUMAN EDGE
SHAPING AGENTIC AI

S ince the initial hype around Generative AI, the technology
sector has accelerated further, fueled by billion-dollar
investments in the ecosystem of hard- and software vendors, and
the physical infrastructure, such as data centers, to deliver AI
models and applications.

CEOs across all industries are recognizing the opportunity and
threat that AI presents to their business and are pushing their
companies to embrace AI. Hiring freezes to delegate more tasks to
AI, reskill-or-exit approaches, and rehiring of employees who were
previously deemed replaceable are impacting morale, as those who
remain need to pick up the workload or quickly find ways to use AI
or face a similar fate.

Under pressure to do more with less, leaders and professionals
use AI-enabled applications to quickly generate, analyze,
summarize, translate, or research information. But creating results

1

faster does not automatically mean they are also better. Instead, generic, low-quality, or incorrect results are passed on to recipients, who must correct them or risk exacerbating the effects if they act on them in good faith.

Users now provide a goal to software that analyzes it, plans, reasons, and acts. Such AI agents act with greater autonomy and automate more complex business tasks than traditional process automation. Adopt the Dual-Lens Principle to manage agents. Treat them like digital employees in operations, and like software in ethics. Operationally, you manage roles, access, and performance. Ethically and legally, the responsibility stays with humans.

When technology becomes more capable and takes on more routine work, the definition of human work and how we measure it needs to evolve alongside it. Depending on the context or profession, it is necessary or recommended to disclose your use of AI, at least in the early phase of technology diffusion. This section introduces you to the accelerating AI adoption in businesses and the need to develop a human edge with AI agents.

CHAPTER ONE

RESISTING SPEED
WITHOUT STANDARDS

S ince its inception in the 1950s, the AI space has seen tremendous optimism about what the technology will achieve and when it will do so. In 1970, Marvin Minsky, one of the founding fathers of the field, was quoted in an interview:

> *"In from three to eight years, we will have a machine with the general intelligence of an average human being."*
>
> —*Marvin Minsky*

Although AI technology has advanced significantly over the past 50–60 years, this vision has not materialized (yet). That does not stop the CEOs of leading AI labs, which develop the foundational models and systems that underpin the latest AI hype, from regularly sharing their predictions for when the industry will reach this level of Artificial General Intelligence (AGI), a state in which AI systems will be as intelligent and capable as humans or even exceed that.

Not surprisingly, there is also a self-serving interest in making such claims, as their companies work toward achieving this capability. The closer this goal is to the point in time at which the claim is made, the more likely the company appears to be on the path to achieving it—or so the implication goes. AI labs also define AGI in their own terms, aligned with their vision and product roadmap, rather than adhering to a universally accepted definition.

Innovation moves quickly, and it accelerates even more in this phase of Generative AI and Agentic AI. OpenAI's release of ChatGPT on November 30, 2022, was a watershed moment for the tech industry, followed by rapid user growth in subsequent quarters. Although many of your friends and family members already use AI regularly, the mass adoption of technology in businesses and society occurs at a much slower pace. Renowned futurist Roy Amara succinctly said in 1978 what is now also known as *Amara's Law*.

> *"We tend to overestimate the effect of a technology in the short run and underestimate the effect in the long run."*
> **—Roy Amara**

How about you? Pause for a moment and ask yourself whether you under- or overestimate the current state of AI's capabilities. Whatever your answer, as a leader or professional, you need to consider the current state of your domain and industry and incorporate new ways of working with AI by your side. Earlier innovation waves have shown that roles will change or disappear.

After all, the occupations of switchboard operators, elevator attendants, and pin boys at your local bowling alley are nearly nonexistent compared to a few decades ago. Changes like that are part of the reality that any professional—current and soon-to-be—faces and *has* faced. There is no playbook for it. The details are still unfolding, but the direction is clear. That makes this period particularly uncomfortable for most people navigating it. Leaders feel the tension between communicating their company's strategy and the uncertainty of this change, or its rationale.

In executive workshops, leaders often ask for a universal playbook or cheat sheet that guarantees reliable results. The desire is understandable. AI introduces variability, and leaders want predictability. But no single playbook exists. Organizations differ in risk tolerance, data access, decision rights, and quality expectations. What does exist is a disciplined approach to define standards, experiment within boundaries, measure rework, and codify what works into repeatable scenarios. Over time, that approach becomes a playbook that fits your organization.

Navigating this unfamiliar territory while working with time constraints can be a natural challenge. Simply reading about AI or consuming information without directly applying it will not work either. Unlike earlier generations of AI technology, the current one requires you to be hands-on to assess where AI could add value and what AI is really capable of at this moment in time. Luckily, it has also never been easier to do so, with capable assistants and apps at your fingertips 24/7.

As a child, I was fascinated by dinosaurs. I assumed their extinction must have happened in a big bang: an asteroid hit Earth, a terrible storm followed, and it got dark and terribly cold. Next, the dinosaurs became part of ancient history. As I now know, that is not how it happened. This change happened over a much longer time, ranging from tens to tens of thousands of years. Figuratively speaking, think of AI as the asteroid that has hit the business world, and the ripple effects are starting to show. That is why both leaders and professionals need to learn and adapt to this new and evolving reality, regardless of the timeframe in which Agentic AI is expected to drive change and its potential impact.

Consider a more recent example, like self-driving electric vehicles (EVs). EVs from automakers like Tesla have been on the market since early 2008. But even nearly two decades after their market introduction, most of us still drive ourselves and get fuel at a gas station. Why is that? The field of innovation theory offers some helpful hints. Visualized as an S-curve in Everett Rogers' research on innovation diffusion[6], or as crossing the infamous chasm[7] after Geoffrey Moore, technology adoption typically follows a pattern of slower adoption by a few early adopters, then mass adoption and scale, ending with even laggards adopting it.

---

[6] Rogers, Everett, 1962, "Diffusion of Innovations".

[7] Moore, Geoffrey, 1991, "Crossing the Chasm".

INNOVATION DIFFUSION

Adoption

→ t

Innovators | Early Adopters | Early Majority | Late Majority | Laggards

CROSSING THE CHASM

Early Market   Mass Market

Adoption

CHASM

→ t

Innovators | Early Adopters | CHASM | Early Majority | Late Majority | Laggards

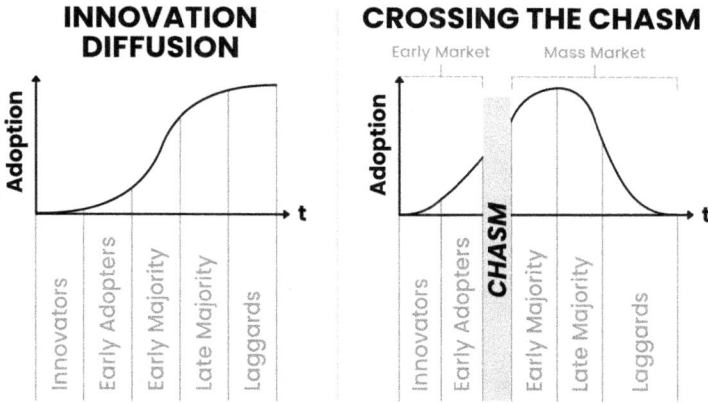

Innovation Diffusion vs. Commercialization Models

Part of the challenge is that innovations are introduced into a status quo environment that cannot be easily changed, or changed in the rapid, radical ways necessary to accelerate mass adoption. For example, it is just not feasible to replace all gas stations and retrofit them with charging stations, or redesign roads for fully autonomous driving. There are economic, legal, societal, and environmental parameters that must be put in place to support it. In most countries, that process takes time.

In October 2015, Tesla introduced its autonomous driving system, but only on highways. Highways are controlled environments where traffic flows one way, lanes are standardized, and vehicles enter and exit via on- and off-ramps. These constraints reduce edge cases compared with open roads, where intersections, pedestrians, and unpredictable driving multiply the variables. Additionally, older cars on the road that lack sensors and similar technologies rely on human drivers (potentially with driving-assistance features such as cruise control). The vision of traffic optimization and making stop-and-go traffic a thing of the past has not materialized yet. But if you want to live in this kind of future and be part of it, you can certainly purchase a vehicle equipped with these features. So, change is happening, and it is possible.

Similarly, AI labs are chasing the headlines. Securing ever-increasing funding rounds and valuations, while cementing their relevance and accelerating growth, are among the top priorities. But there is a significant gap that continues to widen, impacting organizations' ability to adopt these innovations at a rapid pace. Just like the rules of the road, established businesses have processes and procedures to follow—the classic "red tape." Smaller organizations could move swiftly, but they often lack the resources and budgets to keep pace. How can teams use AI responsibly, creatively, and with measurable quality?

Clarifying What Matters as AI Accelerates

Agentic AI is reshaping the workplace. Thus far, though, company leaders have been struggling to find the right tone and to balance internal and external messaging for this change, as it is unclear how exactly it will unfold. As a leader, you need to acknowledge this situation, even if you do not yet have a definitive answer to your team members' questions. However, none of these changes to jobs and roles will occur overnight or simultaneously across all industries and geographies. This should provide some level of comfort, though it would be naïve to assume these shifts will happen only to some future generation or not at all.

The software and technology industry is a good example of what is likely to follow in other industries soon. Once a poster child for stable work, high salaries, and office perks, the technology sector has seen waves of layoffs in recent years, to the point where employees feel the social contract between leaders and labor is broken. Add AI-driven disruption to this mix, and it is unsurprising that fewer employees feel engaged at work.

Driving AI adoption within a company is challenging. I cover the details in the *AI Leadership Handbook*. It primarily involves change management and relinquishing the status quo, even when it is uncertain what exactly the future holds. You should empower employees to challenge the status quo, drive change, and capitalize on the benefits of Agentic AI. Leaders often refer to it as developing an "AI-first" mindset, and this push has led to some rather candid communication from CEOs.

For example, the leaders of software companies *Shopify*, *Duolingo*, *Fiverr*, and *Opendoor* have publicly stated that new roles will only be approved if the requesting manager demonstrates that AI cannot perform the job, contractors will be replaced with AI agents, staff will be reduced, and AI adoption is mandatory; otherwise, individuals will be fired[8,9,10]. The message is simple: Adopt AI or leave[11].

After significant public backlash[12], *Duolingo* walked back and softened their new mantras, including wiping their social media presence and starting anew. Others, like fintech company *Klarna*[13], have realized that pushing for an AI-only workforce to handle the work of 700 customer service agents does not yield the desired results, and they have rehired employees who were previously rendered obsolete. This highlights the delicate balance that leaders must strike both internally and externally when pursuing AI initiatives and seeking operational benefits. On top of that, employee engagement suffers as those remaining will need to carry the additional workload and quickly find ways to use AI themselves.

---

[8] CNBC, 2025, "Shopify CEO says staffers need to prove jobs can't be done by AI before asking for more headcount," April 07, 2025, https://www.cnbc.com/2025/04/07/shopify-ceo-prove-ai-cant-do-jobs-before-asking-for-more-headcount.html.

[9] Kaufmann, Micha, 2025, "Fiver is going back to startup mode," September 15, 2025, https://x.com/michakaufman/status/1967624550020985069.

[10] INC.com, 2025, "'Default to AI or Else,' Says New Opendoor CEO in a Companywide Email. It's a Lesson in Emotional Intelligence," October 01, 2025, https://www.inc.com/justin-bariso/default-to-a-i-or-else-says-new-opendoor-ceo-in-a-companywide-email-its-a-lesson-in-emotional-intelligence/91246527.

[11] The Wall Street Journal, 2025, "The Boss Has a Message: Use AI or You're Fired," November 07, 2025, https://www.wsj.com/tech/ai/ai-work-use-performance-reviews-1e8975df.

[12] The New York Times, 2025, "The C.E.O. of Duolingo Wants to Have a Conversation About A.I.," August 17, 2025, https://www.nytimes.com/2025/08/17/business/duolingo-luis-von-ahn.html.

[13] CNBC, 2025, "Klarna CEO says AI helped company shrink workforce by 40%," May 14, 2025, https://www.cnbc.com/2025/05/14/klarna-ceo-says-ai-helped-company-shrink-workforce-by-40percent.html.

Software vendors are measuring *revenue per employee* as a new metric to determine efficiency and impact. Meta even added AI fluency and quantifiable results for its 2026 performance review cycle[14]. While AI adoption is a key enabler for businesses, most businesses are not even AI-ready. The combination of top-down push and bottom-up experimentation has severe side effects when team members misuse or over-rely on AI tools in good faith. That is why learning how to use AI is so critical. In the immediate term, every leader and professional will need to engage in some learning and experimentation. But using OpenAI ChatGPT, Microsoft Copilot, Anthropic Claude, and similar tools needs to go beyond drafting emails and generic-sounding content quickly. This evolution is not about prompting. It is more about tapping into the potential of augmenting oneself, exploring ideas, confirming findings, role-playing, delegating tasks, and using multiple AI tools, among other things.

Lastly, there is always the realization that entire roles can only partially be automated, and businesses still need humans. While companies are slowing hiring for entry-level roles now, they might realize it is a contraction rather than an elimination of such roles. As a business leader, you need to ask yourself whether you want to accept lowering the quality, even if the result is cheaper to produce. There are plenty of areas in business where a "good enough" result

---

[14] Business Insider, 2025, "Meta is about to start grading workers on their AI skills," November 14, 2025, https://www.businessinsider.com/meta-ai-employee-performance-review-overhaul-2025-11.

is good enough. It is a matter of identifying them in *your* business and managing common risks. You might be wondering how to ensure that the people in your company use AI and remain knowledgeable in their domain. The problem is that many areas in a business cannot afford to create "good enough" or generic information, let alone information that is factually incorrect. Yet not using AI is not an option either.

Exposing Shadow AI and Its Organizational Impact

AI use in businesses is on the rise. As of 2025, Gallup reported that 45% of professionals regularly use AI at work[15], more than double the prior year's level. Yet many companies are just starting to introduce AI into their businesses or have not yet done so. This creates a gap when leaders push for efficiency gains and employees seek ways to make their jobs easier, often using unapproved AI tools that the IT department is unaware of, called *shadow AI.*

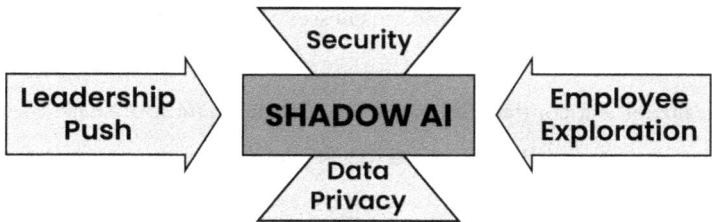

Shadow AI's Impact on Security and Data Privacy

---

[15] Gallup, 2025, "AI Use at Work Rises," December 04, 2025,
https://www.gallup.com/workplace/699689/ai-use-at-work-rises.aspx.

Especially small and medium-sized businesses often lack an official AI policy, and employees are unsure whether and which AI tools they *may*, *should*, or *can* use safely without exposing the company to legal or financial risks. Companies blocking access to AI tools like ChatGPT and Claude on their corporate networks find that employees circumvent these restrictions by entering company data into AI apps on their personal devices. As the CEO of a mid-sized engineering firm shared with me, *"Our IT team says they're only monitoring AI tools within the corporate network."* By taking a selective view on AI use, companies like this are effectively less secure. Turning a blind eye to the reality of grassroots AI adoption does not address or solve the problem.

Slack's 2024 Workforce Index[16] revealed that professionals use AI technology without waiting for formal approval from their managers or IT department, as it helps boost productivity and achieve better results. Nearly half of the participants said they are afraid to tell their managers and peers about their use of AI because it could make them appear lazy or incompetent. A study conducted by Duke University[17], involving 4,500 professionals, found similar results. But here is the thing: it says more about the state of corporate culture than about AI technology if the most innovative

---

[16] Slack, 2024, "The Fall 2024 Workforce Index Shows Executives and Employees Investing in AI, but Uncertainty Holding Back Adoption," November 12, 2024, https://slack.com/blog/news/the-fall-2024-workforce-index-shows-executives-and-employees-investing-in-ai-but-uncertainty-holding-back-adoption.

[17] Reif, Jessica, et al., 2025, "Evidence of a social evaluation penalty for using AI," May 08, 2025, https://doi.org/10.1073/pnas.2426766122.

employees—who are evidently creating new efficiencies with AI—
are afraid to share that they know how to use it (and do so).
Adopting AI well is not a *technology problem*, but a profoundly human
one. Your team members are concerned about how you, as the
leader, perceive them and their performance. That is why your
active guidance as a leader and your efforts to normalize AI use are
so important in driving this change. (We will come back to this
aspect in Chapter Eight.)

Whether you approve of team members using AI tools or
merely acknowledge it, leaving employees to learn how to use tools
like ChatGPT, Copilot, and Claude on their own creates a third
challenge: *over-reliance* and *cognitive offloading*. Team members who are
just beginning to learn about AI tools and agents assume that AI
tools are more capable than they are. It often stems from limited
awareness of and practice with these tools. As a result, team
members rely on AI-generated answers and results when they
should not. There are several well-documented cases in which
lawyers have over-relied on AI without first checking the
information it generated. Over-reliance on AI makes it easier to
perpetuate incorrect information (as we will see in the next section).
You have likely heard the phrase *"AI is not going to replace you, but a
person using AI will."* While it seems that using AI already gives you
an edge over others, it does not when you apply the technology in
the wrong way or context. The key is to develop and retain your
HUMAN Edge by using technology, rather than blindly delegating
to it and trusting it.

Over-reliance on AI often appears as *cognitive offloading*. When teams delegate too much of an end-to-end workflow, they stop engaging deeply with the underlying material. The immediate output arrives quickly, but understanding declines. Over time, that affects judgment, quality, and the ability to detect errors[18].

A useful way to discuss this with your teams is to separate workflows into stages, such as framing, research, synthesis, and final drafting. AI accelerates stages like retrieval and first-pass synthesis. The risk increases when AI owns every stage with minimal human intervention. That is why you should set expectations that preserve learning and judgment, especially in roles where expertise is the product.

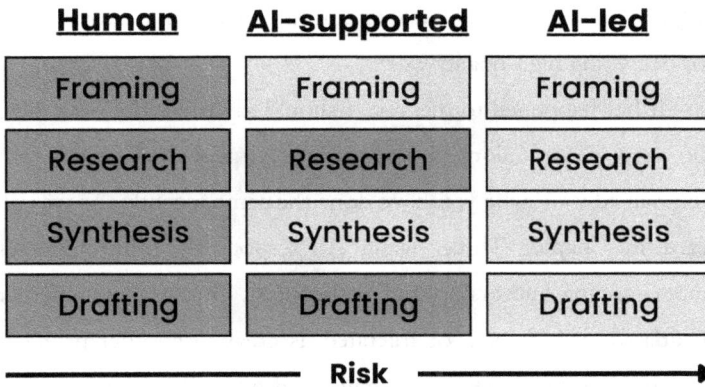

__Human__	__AI-supported__	__AI-led__
Framing	Framing	Framing
Research	Research	Research
Synthesis	Synthesis	Synthesis
Drafting	Drafting	Drafting

◀──────────── **Risk** ────────────▶

Risk of Cognitive Offloading with Increasing AI Support

---

[18] Kosmyna, Natalia, et al., 2025, "Your Brain on ChatGPT: Accumulation of Cognitive Debt when Using an AI Assistant for Essay Writing Task," June 10, 2025, https://arxiv.org/pdf/2506.08872.

Researchers in Switzerland[19] investigating the effects of cognitive offloading on critical thinking concurred that delegating too much to AI decreases the individual's cognitive skills and engagement with new information. This is also critical in business, where the sharpest minds and brightest, most successful ideas attract new customers and generate new revenue for your business.

If you are a leader, you likely spot AI slop in a team member's polished writing, although they have not been a prolific writer before. The telltales include em dashes (—), repetitive words within a paragraph, sentence structures such as *"it's not this; it's that,"* and mediocre content that sounds as if anyone could make the claims and lacks substance and depth. One CEO described it as *"mediocrity, but faster and cheaper."* But it does not mean your team is off the hook for delivering high-quality work.

It has happened to me, too. I should say I have let it happen, too, and passed along a low-quality AI-generated result with minimal edits. Maybe you have done the same recently. As we will learn in Chapter Three, using AI responsibly often involves understanding authorship and authenticity. But AI use, whether mandated, encouraged, or tolerated, is creating another problem for organizations when drafts generated with ChatGPT, Copilot, and similar tools circulate and fill recipients' inboxes, leaving them struggling to keep up and process the information.

---

[19] Gerlich, Michael, 2025, "AI Tools in Society: Impacts on Cognitive Offloading and the Future of Critical Thinking," January 15, 2025, https://doi.org/10.3390/soc15010006.

Preventing AI Workslop While Scaling Productivity

Generative AI and Agentic AI are significant productivity boosters for a range of knowledge work tasks that require deep thinking and specialized skills. Getting to a first draft within a matter of seconds with the help of AI has become an easy decision. Why write it yourself? Generate, skim, and send it. Productivity problem: solved—unless you are on the receiving end.

As a senior manager shared with me: *"Now, I receive all these AI-generated meeting summaries and drafts from my team asking me to review the documents, and I need to spend time going through them."* It is not just that this is AI-generated; often, team members create it quickly and without diligent review. Instead of being more productive because of AI, you actually become slower because of it, as you review and correct mediocre work product.

Imagine the following scenario: You have a broken kitchen stove and hire a professional who comes to your house to inspect and repair it. Before you even say 'hello,' they are already out the door again, sticking a screwdriver and a bill in your hand that they have haphazardly put together.

According to their own account, they are highly efficient because their approach enables them to see ten times as many customers in a workday. The only problem is that your stove is still not fully operational, and you need to troubleshoot and fix the remaining issue yourself. None of us would find this acceptable, let alone pay for such a service.

Yet inside organizations, that same dynamic is spreading. People share AI-generated drafts and call it efficiency, while recipients absorb the burden of verifying it. Instead of "banning AI," treat quality as a contract, and define what is acceptable, what must be verified, and who owns the outcome when AI contributes.

Now, recall the last time a coworker or team member asked you to review a draft they had created. Maybe they even sent you the "final" version, or they shared that they had created it quickly with the help of AI. But once you start reading it, you notice the draft is not bad. It is mainly correct. However, it is neither great, nor specific, nor error-free. In fact, you feel that with a bit more attention and care, the team member could have spotted and resolved these mistakes themselves before sending the draft to you for review. What has just happened? They skipped the Labor-Shift Check, turning their AI-driven productivity gains into your productivity loss during review. While not all AI-generated information automatically creates Draft Debt, it rarely happens just once or with a single team member.

A 2025 study by BetterUp Labs and Stanford Social Media Lab examined this behavior and coined the term *AI workslop*[20]. In this book, we will also refer to it as workslop, AI slop, or simply slop. Merriam-Webster selected the latter as its 2025 word of the year[21],

[20] Perry, Elizabeth, 2025, "The hidden cost of AI 'workslop' — and how leaders can fix it," September 29, 2025, https://www.betterup.com/blog/hidden-costs-workslop.

[21] Merriam-Webster, 2025, "2025 Word of the Year: Slop," December 14, 2025, https://www.merriam-webster.com/wordplay/word-of-the-year.

defining slop as *"digital content of low quality that is produced usually in quantity by means of artificial intelligence."*

Many professionals receiving AI workslop feel annoyed (54%), frustrated (46%), confused (38%), or even offended (22%)—and nearly one third of employees said they would not want to work with the person sending workslop again. Considering these findings, it is no surprise that professionals are reluctant to share their use of AI, as we saw earlier. Reflect for a moment on whether you have recently received a low-quality, AI-generated report or document at work and how you have felt.

Receiving AI slop is not even the worst problem; rather, it is what follows next. Just like the manager in the previous example, team members also spend time reviewing and correcting half-baked, erroneous AI-generated work they receive, which impacts productivity. The estimated cost to correct these mistakes is nearly US$200 per employee per month, and it will likely increase further if this behavior continues beyond this early phase of AI adoption.

While these figures only quantify the additional effort and related costs, the impact is even greater when a product or service your company sells contains AI slop. Take the case of *Deloitte*[22], a global top professional services firm. The company has used AI to generate a report for the Australian Department of Employment and Workplace Relations. The AI model or tool Deloitte used

---

[22] Associated Press, 2025, "Deloitte to partially refund Australian government for report with apparent AI-generated errors," October 07, 2025, https://apnews.com/article/australia-ai-errors-deloitte-ab54858680ffc4ae6555b31c8fb987f3.

produced factual inaccuracies (also known as *hallucinations*). It invented sources that do not exist and misquoted a judge on a matter that government officials will base their decisions on. After the issue became public in October 2025, Deloitte agreed to partially refund the AU$440,000 contract (about US$290,000).

A few weeks later, a similar case in Canada surfaced, in which Deloitte created a *health human resource plan* for the Government of Newfoundland and Labrador that included several references to articles that also do not exist. The impact was nearly CA$1.6 million (US$1.14 million)[23] of taxpayer money.

These issues are particularly peculiar, as the professional services industry prides itself on being "AI-first" while simultaneously facing headwinds over the value and hourly rates of consultants as AI capability and adoption increase. Competitors like *Accenture*[24] have announced plans to "exit" 11,000 employees who cannot be reskilled on AI. At the same time, AI labs like OpenAI claim that AI agents can research information and achieve similar quality for a subscription price of US$20–200 per month.

Although workslop is the latest term for it, mediocre and low-quality results have always existed in business. By using AI, it is becoming even clearer when it happens. AI makes it faster to

---

[23] The Independent, 2025, "Major N.L. healthcare report contains errors likely generated by A.I.," November 22, 2025, https://theindependent.ca/news/lji/major-n-l-healthcare-report-contains-errors-likely-generated-by-a-i.

[24] Fortune, 2025, "Accenture's $865 million reinvention includes saying goodbye to people without the right AI skills," September 27, 2025, https://fortune.com/2025/09/27/accenture-865-million-reinvention-exiting-people-ai-skills.

produce work, teams mark more drafts as "final" (although they are not), review burdens rise, and quality quietly degrades under the weight of volume. That is why leaders must learn to recognize this Slop Spiral. So how do leaders guide hybrid teams of humans and AI agents as the technology advances, empower team members to use AI responsibly, deliver high-quality results with high accountability, and avoid creating AI slop?

Key Takeaways

In this chapter, we have seen how the push for using "more AI" accelerates organizations' internal adoption of AI without proper guidelines and guardrails:

- CEOs are pushing their employees to adopt AI across all business functions. Some leaders promote adopt-or-exit mentalities, causing lasting effects on their workforce.

- AI use is on the rise among professionals year over year, as team members look for ways to increase efficiency and productivity amid rising pressure in organizations.

- Not all AI tools are approved, leading to shadow AI usage that leaves companies vulnerable to security and data privacy risks and dilutes bargaining power with vendors.

- Because of the increase in AI usage combined with limited training and guidelines, more employees are generating and receiving low-quality, generic, and even incorrect information, creating Draft Debt.

- Identifying and correcting slop is cumbersome and often negates the intended productivity gains.
- If AI slop remains undetected, it turns into a Slop Spiral that tarnishes your company's brand and reputation, as early cases indicate.
- Leaders need to guide their teams toward AI readiness and define guidelines for responsible AI use.

CHAPTER TWO

UNDERSTANDING AGENTIC AI
WITH CONFIDENCE

A fter the initial Generative AI scenarios, such as generating, summarizing, and translating text (and other media), a new phase is emerging. This time, the shift is not just automating clicks on a screen (as Robotic Process Automation does), completing individual approval steps based on patterns in data (as Machine Learning does), or working with language (as Large Language Models do).

This phase uses software "agents" that can plan and execute multi-step work toward a goal, often across tools and systems. In practice, this phase focuses on using agents to automate tasks with limited uncertainty and complexity, while keeping humans in control of direction, approvals, and outcomes. The result is faster completion of repetitive knowledge work and more consistent execution across teams and processes.

Introducing Agentic AI as a New Operating Model

AI agents represent the next level of software for automating business processes and making decisions under uncertainty. They often act as interactive personal assistants within business applications, including productivity tools such as Microsoft Office, customer relationship management (CRM) systems such as Salesforce, and procurement and finance applications from vendors such as SAP.

For example, a human customer service agent uses an AI agent to summarize recent customer interactions and case history. AI agents also interact with each other to complete tasks, such as a customer service agent and a reviewer agent who offers feedback on responses. Several principles of organizing human labor also apply to agents, as we will see in Chapter Six.

The top business benefits of AI agents include:

- Managing routine tasks that still involve some uncertainty, so team members can focus on higher-impact priorities. For example, an agent summarizes a meeting, identifies action items, and distributes a draft recap to stakeholders.
- Quickly analyzing large datasets and recommending next steps, so teams can review and decide instead of starting from scratch. For example, agents connected to business systems (e.g., Salesforce CRM or cloud file-hosting services like Dropbox) review sales data by category, propose account strategies, and draft customer outreach.

- Tailoring information for customers or users to support more personalized communication. For example, an AI-enabled customer service agent retrieves relevant product information and drafts a response that reflects purchase and support history, preferences, and a preferred tone, accelerating resolution and strengthening brand loyalty.

To better understand AI agents and the relationship between humans and them, the concept of agency is key. Agency is the capacity to act toward an outcome within an environment, and it involves acting independently, making decisions, and influencing the world the agent interacts with. Humans also intentionally delegate a task and the necessary authority from one entity to another to achieve a desired outcome while staying accountable. For example, you delegate agency when you ask a tax professional to prepare and send the annual income tax return on your behalf, or when you hire an attorney to represent your legal interests.

Delegating agency typically follows a common set of steps:
1. Identify the objective or responsibility.
2. Determine the context.
3. Provide data and authority (collaboration).
4. Define expectations and intended outcomes.
5. Monitor progress toward the deadline.
6. Acknowledge and evaluate results.

Humans not only delegate agency to other people but also to software to complete tasks and automate business processes. Traditionally, software has automated straightforward tasks using clear rules and instructions (*if this happens, then do that*). However, real-world situations are often more complex than traditional software addresses. AI agents enable the automation of more complex tasks by allowing users to specify a goal rather than providing detailed step-by-step guidance.

A business strategist asks an agent: *"Research market trends for lawnmower manufacturers in the USA from the past two years, focusing on commercial products."* The agent determines how to best achieve the goal, breaking it down into tasks and steps that it completes one at a time. Like human assistants, AI agents request additional information from the user and use other specialized AI agents to gain expert insights for specific tasks, thereby increasing their versatility and utility.

Consider a quarterly account review for a strategic customer. Instead of gathering information from customer relationship management entries, emails, and past presentations, you provide essential details: the customer's name, timeframe, and purpose of the review, and delegate the task to a team of specialized agents. One agent retrieves recent sales activity and opportunity data. Another compiles support tickets, service issues, and customer satisfaction trends. A third one analyzes product usage, contract terms, and upcoming renewals. A fourth agent reviews pricing rules and discount policies to ensure recommendations remain compliant. Together, they assemble a complete view of the relationship. Once the data is gathered, the agents evaluate it for shifts in order volume, declining engagement, emerging risks, or new areas of demand. They surface the insights that matter and prepare a structured account brief with a customer snapshot, performance highlights, potential concerns, and growth opportunities. They also outline talking points for your meeting.

As you have been working with the customer for a while, you have a personal relationship and additional context that your team of agents does not have. You confirm the narrative, adjust the interpretation, and ensure the recommendations align with your strategic goals and the relationship. Your AI agents save time by gathering and analyzing data, and you review and refine the results to ensure they are accurate and useful for achieving the goal.

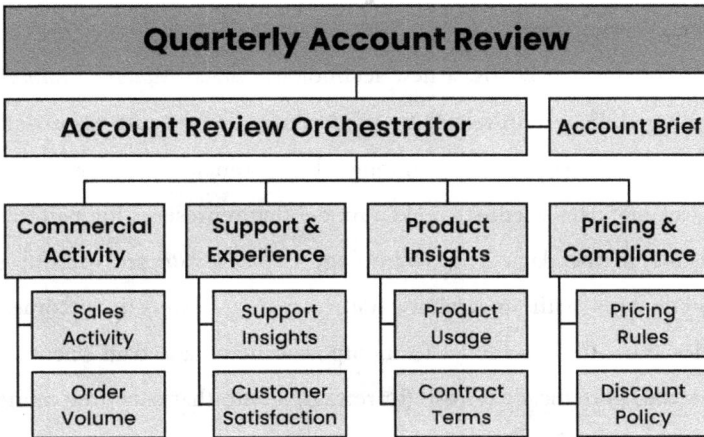

Orchestrating Quarterly Account Reviews with a Team of Agents

Since the 1960s, organizations have used software to automate business processes across finance, human resources, procurement, sales, and customer service. These applications record transactions, streamline communication, and support forecasting and reporting. Computer scientists have traditionally programmed software using rules. Given the same input, users receive the same output every time they run the application. This type of behavior is *deterministic*.

A software engineer explicitly defines logic and reproduces the outcomes if something is not working as expected. But adapting to changing conditions requires changing the application's code. If your code and logic are complex, changes and testing for unintended side effects become time-consuming and costly.

Take a common business process, such as invoice-to-cash, in which you send invoices to customers and receive payment for your products or services. This process typically follows the same rules—every time.

In the mid-2010s, a new technology, called *Machine Learning*, emerged. Instead of relying on rigid rules, it detects patterns in data and learns correlations among data points. Using Machine Learning, data scientists build a model that represents the patterns found in the data. This technology is *probabilistic* and produces predictions with uncertainty, rather than guaranteed outcomes. However, for the model to incorporate new data, data scientists need to periodically refresh (or retrain) it with that data. Retraining requires significant effort, computing resources, and verifying that the updated model works as intended.

Take customer service, for example. By analyzing hundreds of service tickets, organizations have determined the most likely category of incoming requests based on their content: password resets, new applications, or requests for marketing collateral. This speeds up ticket classification and routes the request to the right team or subject-matter expert who addresses the question.

By the early 2020s, *Generative AI* had become the next AI evolution to make headlines, with ChatGPT being the most recognized example of this category of AI-enabled tools. Like Machine Learning before it, Generative AI is also probabilistic in nature. Generative AI models are trained on examples of data such as text, images, audio, and video, and allow users to generate media across these formats and to analyze, summarize, or translate (or transfer) information.

Use your favorite AI assistant to explore how Generative AI and Agentic AI improve work in your role and industry. You will find a sample prompt in the appendix that you can adapt to get started. The generated output will look similar to the one in the table below and will indicate 40 potential opportunities for using AI in your business function, based on the input data and operations. Use it as a directional starting point to further explore and validate concrete ideas.

Prioritize opportunities that clearly link to measurable business outcomes, such as productivity gains, quality improvements, risk reduction, or faster decision-making.

Input	Operation	Output	Example
Text	Transfer	Text	• Translation • Style transfer • Software code
Text	Generate	Image	• Image generation • Concept mockups
Video	Analyze	Text	• Insights • Communication coaching
Audio	Summarize	Text	• Meeting minutes • Call summaries

Examples of Generative AI Across Modalities

Toward the end of 2024, commercial software vendors began delivering the first AI agents within their applications and platforms as part of Agentic AI, a new category of AI technology. What makes this technology stand out is its ability to comprehend an abstract goal, break it down into subgoals, assess potential options for achieving those subgoals, plan and execute the necessary steps to accomplish them, using additional tools and systems if needed.

Agents also evaluate how well they have achieved the original goal and dynamically adjust their approach to optimize the outcome, adding resilience without explicitly defining it. This increases automation and scope compared to earlier technologies.

Let's explore what this looks like in customer service. Human agents traditionally review incoming customer inquiries, check the conversation history, and gather supporting information. Repeating these steps becomes tedious and extends response times. During peak periods such as holidays, the only practical way to scale has been to add staff.

Augmented with Agentic AI, a customer service team member instructs their AI agent to collect relevant customer information. The AI agent evaluates the request, assesses the tone, reviews historical interactions and recent purchases, and drafts a response. Additionally, the AI agent retrieves data from multiple documents and systems, including CRM, and integrates the information into its reply. Once the agent has prepared the draft, the human team member reviews and refines it before sharing it with the customer. While the AI agent gathers and drafts responses, the customer service representative focuses on more complex cases or engages directly with customers, thereby fostering stronger relationships between clients and the organization.

AI agents have three key characteristics across business functions:

1. Automating tasks involving several steps by analyzing and following goals, rather than explicit instructions.
2. Dividing goals into manageable subgoals and tasks to streamline oversight and execution of the process.
3. Synthesizing and analyzing data from disparate systems and documentation to generate informed responses.

This book adopts the following distinction between Artificial Intelligence, Machine Learning, Generative AI, and Agentic AI:

- Artificial Intelligence is the broader domain and software capability.

- Machine Learning is a sub-discipline of AI focused on recognizing patterns in data.

- Generative AI is a sub-discipline of Machine Learning that generates information based on examples.

- Agentic AI is a sub-discipline of Generative AI and enables goal-driven systems.

Whenever you see the terms Machine Learning, Generative AI, or Agentic AI, they refer to specific technologies, whereas the term AI refers to the broader field, encompassing all technologies under it, such as Machine Learning, Generative AI, and Agentic AI.

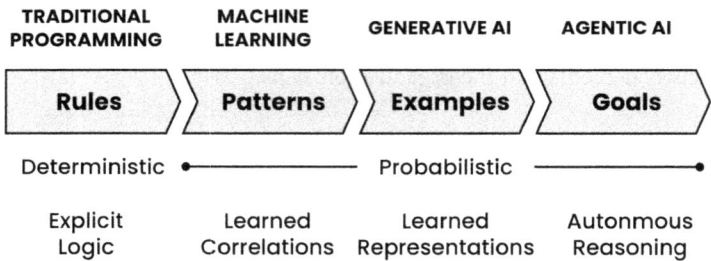

TRADITIONAL PROGRAMMING	MACHINE LEARNING	GENERATIVE AI	AGENTIC AI
Rules	Patterns	Examples	Goals

Deterministic •————— Probabilistic ————————•

| Explicit Logic | Learned Correlations | Learned Representations | Autonmous Reasoning |

Characteristics of Artificial Intelligence Technologies

Businesses are accelerating their use of AI agents to perform the *same task (scope)* as human workers, at a *fraction of the cost,* in a *fraction of the time,* and with similar risks. Unlike the well-known iron triangle, which dictates that a product cannot be comprehensive, cheap, and built quickly at the same time, Agentic AI enables a *golden triangle* that is much more malleable. Keeping one variable steady positively affects the other two. For example, agents can deliver the same scope faster and at a lower cost, or more scope in the same timeframe and at the same cost.

Finance teams use agents to create monthly management reports in hours instead of days by gathering and analyzing data, drafting narratives, and reducing costs while shifting human effort from manual preparation to validation, judgment, and oversight.

Keep in mind that quality requires governance, speed increases the risk of unverified results, and cost savings shift effort from creation to review.

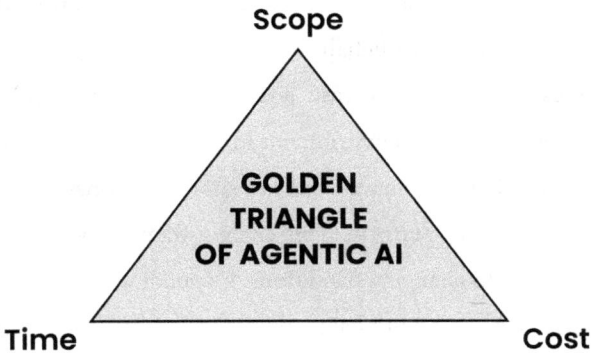

Scope

**GOLDEN
TRIANGLE
OF AGENTIC AI**

Time   **Cost**

Agentic AI's Influence on Project Variables

Defining AI Agents and Their Operational Scope

Most tasks in a business are knowledge-driven. Whether it is knowledge of *where* to find information (location), *who* to talk to (contact), or *how* to perform a particular task (process), knowledge relies on the information and data on which it is based. Agentic AI has accelerated since the release of the first LLM-based agent frameworks a few years ago.

AI agents are software components that make decisions under uncertainty based on defined goals and interact with their environment. The concept of agents is not particularly new, though, and it has existed for decades. For example, the thermostat in your house is an agent. A sensor measures the current room temperature, and if the measured temperature is lower than the value you have set, the thermostat activates the heating until the room reaches the target temperature.

Software vendors build agents into applications and make them available as stand-alone extensible frameworks. Agents execute tasks on a user's behalf, which is a valuable proposition for businesses seeking to increase automation in their operations. Agents understand the environment in which they operate and its current state. They also have access to short- and long-term *memory* and *plan* the task to perform. This enables them to work toward a goal and adapt to changing conditions. Depending on the setup, AI agents use *tools* and additional resources to achieve their goals, for example, business systems and data.

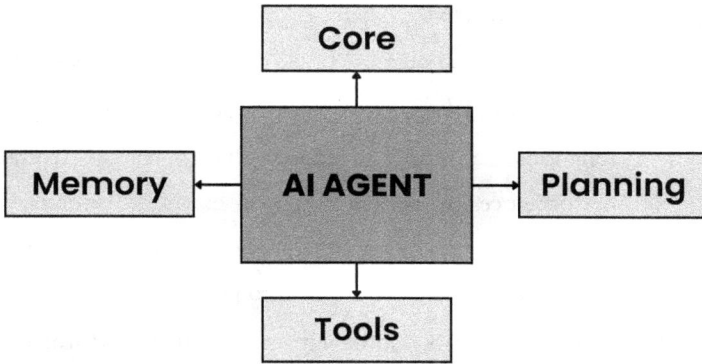

Key Components of AI Agents

In business, agents perform desk research, create or respond to requests for proposal (RFP), or write software code. Depending upon their type, agents process information with an increasing scope while perceiving and interacting with their environment:

- **Simple reflex agents** operate by directly mapping situations to actions, making them suitable for environments with clear cause-and-effect relationships, such as a thermostat or basic e-mail filtering based on user-defined "if-then-else" conditions.

- **Model-based agents** maintain an internal model of the world, which enables them to plan and adapt more effectively to changes by simulating multiple future scenarios and conditions for risk assessment.

- **Goal-based agents** prioritize actions that lead to desired outcomes. This behavior is ideal for tasks with well-defined objectives, such as determining the optimal reorder time and quantity based on market and sales data.

- **Utility-based agents** evaluate actions using utility functions and make decisions that maximize overall performance or satisfaction, for example, optimizing pricing strategies to maximize revenue while considering inventory levels and willingness to pay.

- **Learning agents** continually improve their performance over time through experience, making them particularly valuable in dynamic, unpredictable environments for detecting fraud and identifying evolving anomaly and suspicious transaction types.

- **Hierarchical agents** use multiple levels of agents to manage complex tasks efficiently, making them well-suited to intricate systems with multiple subsystems. In sales preparation, agents collaborate to compile a customized pitch based on historic transactions, preferences, financial data, market data, and white spaces.

This last type of agent is especially relevant when tasks become more complex and span agents across multiple domains. Similar to human teams, there are several key roles in these *multi-agent systems* as well that are important to be aware of. The following three roles represent the core concept of agent roles. *Worker* agents perform highly specialized tasks, such as researching customer information or drafting responses to customer inquiries. A *reviewer* agent analyzes the output generated by a *worker* agent and determines whether it meets the system designer's objectives. In customer service, these goals could include readability, relatability, and accuracy. If the output does not meet the expected level for these criteria, the *worker* agent is instructed to gather additional information and regenerate the output for further review. An overarching *orchestrator* agent manages the entire workflow, dispatches tasks to the appropriate worker agents, and ensures the agents deliver verified, accurate results.

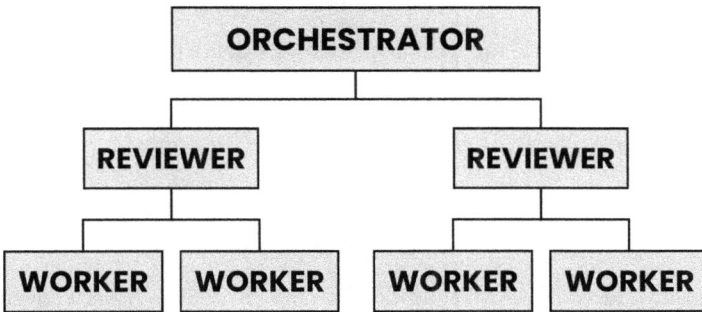

Common Roles of AI Agents

Software and technologies are abundant on the market today, but not all of them are truly agentic. Traditional workflow software enables IT teams and professionals to model the steps in business processes, such as password resets or purchase order approvals, typically following predefined rules. Other examples, like RPA, help to automate repetitive parts of a business process based on predefined steps:

1. Log in to your company's vendor invoice portal.
2. Locate the invoices received within the last 24 hours.
3. Download this subset of invoices.
4. Extract information from these invoices.
5. Enter this information into your finance application and save it.

These examples follow rules without any reasoning or sophisticated decision-making under uncertainty. Hence, they are not agentic. Unlike AI agents, *agentic workflows* leverage Agentic AI capabilities at individual steps. Agentic capabilities complete one or more tasks in a workflow, but the workflow follows a defined pattern.

Applying Systems Thinking Beyond Point Solutions

Just as team members do not work in isolation, AI agents function within a network of interactions involving people, tasks, tools, and other agents. Leaders aiming to deploy AI effectively should consider the whole system or process, not just individual tasks and outcomes. Therefore, leaders need to understand how workflows move across team members, departments, customers, and suppliers. Additionally, understanding what information is needed, where and how to obtain it, and what outcomes are expected is essential. Access to business data across functions such as finance and procurement is crucial. It is also important to identify potential bottlenecks, rework, or misunderstandings in these areas.

Think about Agentic AI in three dimensions: *individual productivity, operational efficiency, and strategic differentiation.* Agents increase team members' *productivity* by conducting desk research on their behalf or by augmenting tasks that humans currently perform. These opportunities are specialized and often limited to an individual's scope of work.

As part of the *operational efficiency* dimension, businesses use AI agents to improve core processes, including cash collection, dispute resolution, interview scheduling, and workforce analysis. *Strategic differentiation* allows your business to deliver benefits to your customers in ways that set it apart.

Consider improving customer satisfaction by introducing AI agents into customer support. The agents handle incoming questions within minutes rather than hours, and human agents remain involved as reviewers. This collaboration provides greater autonomy as your team's confidence in the technology increases.

STRATEGIC
DIFFERENTIATION

OPERATIONAL EFFICIENCY

INDIVIDUAL PRODUCTIVITY

Expanding Agentic AI's Business Value and Impact

For example, in software engineering, a team of agents takes on specialized roles, such as an *analyst, developer, tester, and reviewer.* The *analyst* plans a feature and drafts the specifications and requirements based on a set of goals and instructions provided by a human product manager. A *developer* analyzes these specifications and writes code, while a *tester* creates test plans and reviews the generated code for errors. Finally, a *reviewer* agent looks for any issues that the earlier agents have missed and ensures adherence to style and security standards. Together, this digital team lets human developers scale their work.

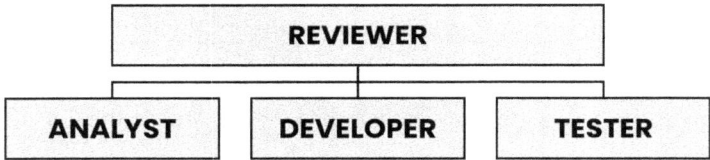

Common Roles on Agentic Software Engineering Teams

The idea of creating a virtual team of agents also applies to any other business function. In marketing, a team of agents creates a campaign by identifying the ideal customer profile, using information about your company's product or service, and incorporating style guides, messaging, and creative assets. These five points will become second nature when building an advertising strategy and buying and placing online ads[25].

---

[25] The Wall Street Journal, 2025, "Meta Aims to Fully Automate Ad Creation Using AI," June 02, 2025, https://www.wsj.com/tech/ai/meta-aims-to-fully-automate-ad-creation-using-ai-7d82e249.

Take a moment to consider your own business function:

- Where do multiple team members with different roles and skill sets work together to complete a task or project?
- What skills are needed for each role, including planning, reflection, implementation, and review?
- Which of these tasks depend on available information and documented procedures?

This is where systems thinking plays a crucial role. Instead of focusing on individual tasks AI handles, elevate the perspective by asking how AI agents integrate into the entire process. Just like in human collaboration, it is important to clearly define the information agents need and generate, set expectations for quality, and establish guidelines. To achieve this alignment, engage peer leaders from business functions your team frequently interacts with. Talk about how you are currently using or plan to use AI agents, and coordinate with their initiatives to ensure smooth collaboration when AI agents are involved.

Using AI agents effectively requires regular reflection and iteration. Structured assessments and activities help you track progress, reinforce best practices, and identify areas for improvement. Review the agents you use, the workflows they support, and the outcomes they deliver. Ask your team for input on whether the generated output still meets their expectations, whether the quality has remained consistent or declined, and whether any data sources have changed. These reviews help identify issues early and ensure reliable results. Also, look for manual steps in your workflow. AI capabilities advance quickly, and you now automate tasks that seemed impossible a few months or quarters ago. As your teams' confidence in AI agents grows, consider expanding your agents' responsibilities by assigning more tasks or integrating additional tools and systems.

Moreover, do not rely solely on the tools your company provides; actively seek opportunities to use Agentic AI in a familiar task or process. Review your process to improve efficiency with Agentic AI, and use agents for research, decision-making, and taking action. Identify existing AI agents or features your vendor or IT team offers to combine with your expertise and techniques for strategic AI use.

Agents are already changing how work is done and by whom. The first companies are experimenting with agents to handle more complex tasks, such as answering customer service inquiries for common categories or hyperpersonalizing the user experience for personal investors. Beyond individual agents, two or more agents also collaborate, for example, to answer a broader range of customer questions across multiple product lines and verify answers before sending them to the customer. Yet even expanding this concept to agents working across multiple departments is no longer just a vision, as customer service and finance agents collaborate to verify a credit note and respond to a customer with the relevant information.

Although most of the coverage of using AI in business focuses on working with text, you should explore broader concepts and ideas as well to determine potential applications in your business. This is also known as *multi-modal* AI scenarios.

Imagine your business manufactures hand mixers. A user interacts with the *customer support (AI) agent* on your company's website. The user uploads an image of their mixer that has stopped working and provides a short problem description. The agent requests a specialized AI agent for visual search to identify the model. Once it has retrieved the model information, the agent sends it back to the first agent, who then passes it to a third agent specialized in identifying and articulating troubleshooting steps, to eventually provide the recommended steps back to the user. The user follows the steps to get their hand mixer working again.

Envision a different example. Your *product marketing* team wants to understand how their closest competitor communicates their product through publicly available videos. Pointing the AI agent to a set of YouTube videos, it processes the transcripts, analyzes the style and tone, compiles a summary of the anticipated target audience, and explains why the chosen phrases and examples resonate. Taking it a step further, the AI agent proposes effective strategies to counter the positioning, for example, in spoken language, as a marketing consultant would.

Multi-modal models, such as OpenAI's GPT or Google's Gemini families, enable AI agents to perceive and interact with their environment in even richer ways and dimensions beyond text.

In the next chapter, we will review the human factors in the use of Agentic AI and how roles and the definition of work change.

Key Takeaways

In this chapter, we introduced key terminology that prepares us for the concepts we will discuss in the following chapters:

- Agentic AI is the latest evolution of AI, enabling users to define a goal for an agent to work toward.

- AI agents use LLMs for planning and reasoning. Agents also have short- and long-term memory functions to maintain current and historical information.

- Agents assume dedicated roles in a business process. Access to relevant information within the scope of their tasks, such as finance, procurement, or CRM systems, enables agents to produce relevant and contextual results.

- Agentic workflows incorporate Agentic AI capabilities at specific steps. Unlike AI agents, they depend on a structured workflow.

- Although agents work on a single task, a team of specialized agents can also solve more complex tasks.

CHAPTER THREE

REDEFINING HUMAN VALUE IN HYBRID TEAMS

Modern management practices evolved in environments where human capacity was the constraint for work. Specialization, division of labor, and visible effort became reliable proxies for value. Many organizations still reward activity and output volume because those signals were historically correlated with contribution.

Agentic AI disrupts those proxies. When software completes parts of knowledge work rapidly, effort becomes harder to observe and less useful as a measure of impact. Leaders need a clearer definition of value that does not depend on visible busyness. The bottleneck shifts from execution to judgment, framing the right questions, selecting the right approach, validating outcomes, and taking responsibility for decisions. In day-to-day operations, many employees lack the time and the permission to increase efficiency, and they struggle to follow the mantra *"work smarter, not harder."*

AI agents make this gap more transparent. It is tempting to view agents as equivalent to humans for narrow tasks, but the analogy breaks at the role level.

Deciding When to Treat AI Agents as Digital Employees

It is easy to compare agents to humans or refer to them as *digital employees* who analyze, plan, reason, and act, or suggest stopping hiring humans altogether[26]. But agents are still software components that work on well-defined domain problems. Yet there are instances where drawing parallels between humans and agents is helpful, which we will discuss further.

Compare human workers and agents across six aspects of the HUMAN Agentic AI Edge *Operating Model*™ (roles, knowledge, rules, rewards, collaboration, and organization) to help you design delegation boundaries and define accountability, but do not imply that agents possess judgment or responsibility.

1. **Roles:** Agents are taking on specialized roles. This means a marketing agent will dispatch requests to the appropriate subordinate-level agent that specializes in creating ideal customer profiles, search engine optimization (SEO), or web copy. This approach is similar to dividing work among human managers, teams, *and* subject-matter

---

[26] TechCrunch, 2025, "Artisan, the 'stop hiring humans' AI agent startup, raises $25M — and is still hiring humans," April 09, 2025, https://techcrunch.com/2025/04/09/artisan-the-stop-hiring-humans-ai-agent-startup-raises-25m-and-is-still-hiring-humans.

experts. Just like the scope of roles in a business is defined through job descriptions, AI agents also need a definition of their specialization, knowledge level, collaborators, authority, and available systems and access.

2. **Knowledge:** Agents must keep their knowledge current to be valuable to the organization. That means they need access to the latest and most accurate data to inform their decisions. While humans initially provide new data, other agents could eventually take over this task. These agents act once they notice that the first agent is not meeting the user's or customer's expectations. Think of how internal communications and HR teams help to keep team members' knowledge current today.

3. **Rules:** Agents also need to abide by the rules and regulations of their specialization, such as how to close the books in accounting or how to comply with regulations when recommending solutions. Before an AI agent performs any task in your business, it will need to ingest, understand, and accept the boundaries of acceptable and unacceptable behavior, like a code of ethics and a code of conduct. Think of it as the key steps for a new hire on the first day and a periodic recertification.

4. **Rewards:** Agents have a built-in reward function that provides feedback about how well they achieve their goal. Because agents maximize their reward, you should regularly review their goal attainment and the results they

achieve, just as you would in a quarterly performance review or check-in. To assess your agents effectively, ask your team to document cases where the agent did not deliver the expected results, and gather context on the instructions and any data provided to the agent. As similar incidents trickle in, determine if the agent needs to be updated or if your team needs to refine its instructions.

5. **Collaboration:** Few agents will work in isolation. Instead, they collaborate with other agents to accomplish a goal. Take the earlier example of the marketing team. Agents must communicate with other agents to understand the goal, share interim steps, gather feedback, and report the results. But that is not enough by itself. Just as your organization's leaders have defined a shared vision and mission for the company and team, agent developers need to ensure that every agent involved in achieving a business goal also shares that goal. This will increase the likelihood of achieving the best possible result for the company.

6. **Organization:** If tasks require input from other departments, such as requesting a budget increase from finance or publishing web updates by IT, governance, and awareness of organizational structure, teams, and other agents' specializations are key. Organizational charts and internal address books are the equivalents of this idea today to find out who works in this business, what their role and reporting line are, and what they help with.

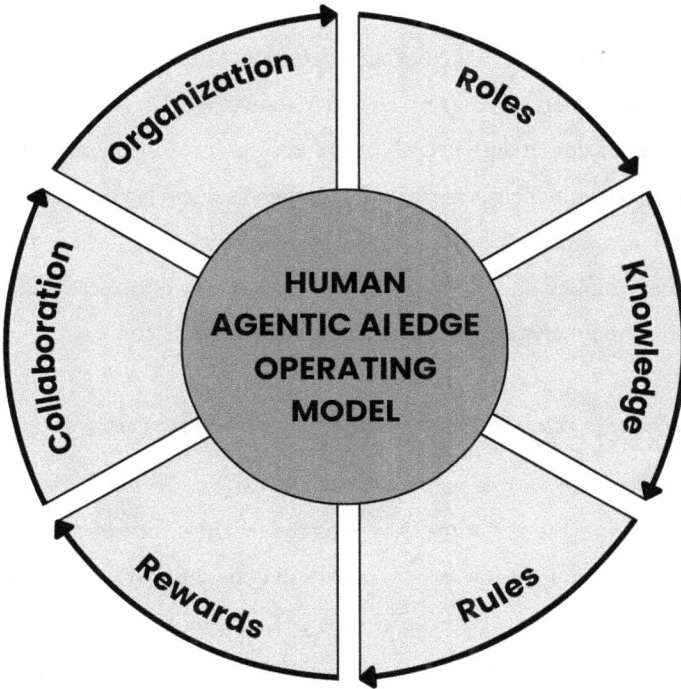

The HUMAN Agentic AI Edge Operating Model™

With this operating model in mind, apply the Dual-Lens Principle of operations and ethics. Manage agents like employees, drawing on established concepts of organizing human work. (We will discuss how common HR practices and processes extend to Agentic AI in Chapter Seven.) Govern agents like software, while humans remain ethically accountable for the agents' actions.

Despite the human-like capabilities of Agentic AI, leaders also need to understand when *not* to compare agentic and human labor.

Although AI agents process vast amounts of data quickly, they operate only on the data they have access to. Human team members have additional information and tacit knowledge that is not digitally captured, documented (well), or accessible by AI. Agents have a defined and highly specialized scope, such as resolving credit blocks on a customer account. But the human finance professionals in your company doing this work today have a much broader skill set and scope than a single task, making it an unequal comparison.

Reframing Human Work in an Agentic Environment

From an early age, we are praised for our accomplishments, from helping with chores to kicking the ball into the net or getting good grades. In many ways, work—and completing it—is a natural progression that helps professionals feel a sense of accomplishment and receive a monetary reward. AI threatens that mental model with its rapid pace and broad scope of capabilities. As AI augments human capabilities, you and your team members will no longer have to complete all tasks and steps yourself to achieve an outcome.

In this wave of Agentic AI, the technology already completes several tasks that are part of many white-collar roles that require advanced degrees. These tasks involve gathering data, analyzing it, developing options, and making recommendations about the best course of action. However, the roles humans hold are more than the sum of individual tasks, spanning industry experience, personal relationships, and tacit knowledge. Remember that AI agents operate at a narrower task-level rather than a broader role level.

Leaders preparing their teams for the changing workplace act as role models for the following ideas:

- **Work shifts from effort to** *purpose.*
  Clarify who the team serves when completing a task or project. This elevates the discussion from task-level responsibilities to the team's impact on its customers. What professionals give up when they delegate individual tasks to AI becomes easier if the mission is clear.

- **Outcomes shift from knowledge to** *quality and speed.*
  Team members measuring their personal value through task completion and knowledge will instead be measured on delivering results quickly and with high quality. Support team members in assessing and fostering their strengths, and clarify the tasks that employees should spend their time on in a project to achieve the desired outcome.

- **Rewards shift from accomplishment to** *impact.*
  Rather than just rewarding the accomplishment of a goal or task, highlight the actual impact it enables, such as revenue, cost savings, or customer satisfaction.

The shift toward these measures prepares your teams to feel less threatened by technology, as their unique skills for achieving a business outcome become more relevant and valued. Managing team dynamics like that becomes even more relevant as your business rolls out AI agents, with tasks taking minutes or seconds rather than days or hours. Just as leaders are no longer responsible for task-level outcomes every step of the way, professionals are also becoming leaders of sorts—of AI agents.

Dimension	Present	Future
WORK	Effort	Purpose
OUTCOMES	Knowledge	Quality & Speed
REWARDS	Accomplishment	Impact

Evolving Dimensions of Human Value at Work

AI is changing how your teams work and how your company delivers value to its customers. Eventually, this shift will affect everyone and every role. Yet companies measure expectations and experiences against standards that predate the mass adoption of AI. This is a trap that applies to managers and employees alike. It is not trivial to envision the future of work. But when technology evolves, so should our definition of work. One of these changing aspects is whether and when to disclose that you use AI.

Building Trust Through Transparency

Successful leaders of both human and Human-Agentic AI teams are aware that any message they share conveys four aspects (fact, self-disclosure, social relationship, and wish). German psychologist Friedemann Schulz von Thun first described them in his *four-sides model*[27]. The recipient of a message interprets it based on one of these sides. A mismatch of sides between the sender and receiver is the most common reason for misunderstandings in any communication. For example, your significant other tells you, *"Is it cold in here?"* nudging you to close the window (self-disclosure). If you interpret it as a fact and respond with *"No, it's fine,"* you have a classic misunderstanding.

Depending on the context in which you communicate, the recipient has different expectations about what they receive from you, including the level of personalization and authenticity.

The question of whether to have a ghostwriter or speechwriter author your book or speech, versus the authenticity of writing yourself, is a challenge that only individuals with access to professional writers have had to ponder. But as anyone can now access personal AI-powered ghostwriters, it is important to consider when to use AI to generate information.

A key reason why AI workslop is proliferating in businesses is the mismatch of authorship and attribution. In 2023, one of the

---

[27] Schulz von Thun, Friedemann, 1981, "Miteinander reden: Störungen und Klärungen. Psychologie der zwischenmenschlichen Kommunikation".

first widely reported cases of ChatGPT misuse appeared in the media. Two lawyers in New York City had used it to search for precedent cases they cited in documents sent to the court[28]. But several of these cases did not exist.

The court found that ChatGPT had fabricated several of the cited cases, and the lawyers using the tool were unaware that the underlying Large Language Models (LLMs) merely predict the next word in a sentence and do not verify the accuracy of their output. Although it was a heavily publicized case, fellow lawyers have fallen into the same trap even two years later[29].

But it is not just lawyers who have a special duty to be diligent and to work based on accurate information. The same is true in any other profession. When deciding whether to disclose the use of AI, consider both the nature of the work and the recipient's expectations.

For example, clients consult and pay experts in research, academia, and professional services for their unique and authentic expertise, not for an LLM's output. If you are faced with this situation, ask yourself if someone is expecting a work product because of your role, expertise, or relationship.

---

[28] Reuters, 2023, "A lawyer used ChatGPT to cite bogus cases. What are the ethics?," May 30, 2023, https://www.reuters.com/legal/transactional/lawyer-used-chatgpt-cite-bogus-cases-what-are-ethics-2023-05-30.

[29] Reuters, 2025, "Lawyers face sanctions for citing fake cases with AI, warns UK judge," June 06, 2025, https://www.reuters.com/world/uk/lawyers-face-sanctions-citing-fake-cases-with-ai-warns-uk-judge-2025-06-06.

If you answer with yes, take extra care when creating and reviewing AI-generated output. The two dimensions of Schulz von Thun's communication model at play are fact and social relationship. While the information (fact) you create is correct and meets your quality expectations, the recipient also looks for the second dimension (social relationship) to meet their expectations. This is especially true when the recipient expects you to respond personally, or when work is directly attributed to you as the author, creator, or inventor.

	Institutional	Personal
Authorship	Attributed to company	Directly attributed to author
Guidance	General information	Personal or professional advice
Rigor	Brand voice, tone, and style	Evidence-based analysis
Human Quality	Accuracy, "good enough"	Empathy, authenticity

Dimensions of Communication

Other situations allow for more organizational distance, such as when the work reflects the company rather than a person, provides general information, follows brand guidelines for voice and tone, or only needs to be correct ("good enough") rather than also personal. Whether the recipient expects personal authorship and connection or organizational clarity and consistency from an institution, company, or team, guides when to disclose your AI use.

Spotting when AI-generated output meets the quality level the task demands is typically easier with increasing experience in a role. Over time, leaders and professionals have built a rubric (or frame of reference) for what good results look like in their business function. Sometimes, these criteria are absolute, like a screw and a nut fitting together, or being too loose. Other times, they are much more subjective, like tailoring a report for a target audience. The more precise the criteria are for what is considered high-quality results, the easier it is for professionals to adhere to them. In addition to defining what *good* means, it is important to be clear about the bar for *good-enough* results.

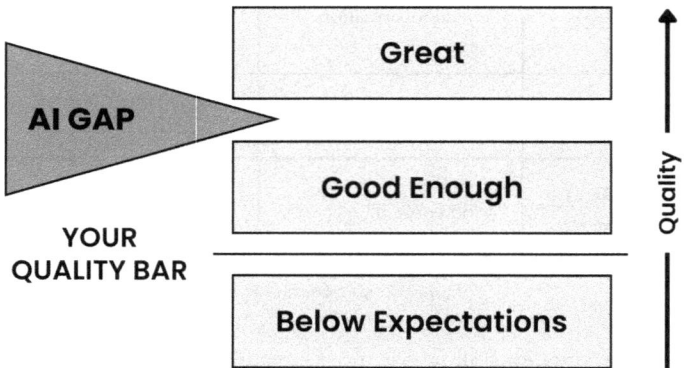

Description of Quality Expectations

For example, when summarizing meeting discussions and sending out action items from project meetings, good enough accuracy and completeness of the AI-generated information is often sufficient. If the project team is aware that you use an AI agent to summarize and send this information afterward, it is rarely a problem if the agent does not get the tone right. Picture the same scenario for your meetings with your company's board of directors. Here, you will want to be even more diligent as you are sending it in your name or on behalf of the executive team, and the decisions communicated in the summaries have a significant impact on the business. As you can see, it depends on the task and the recipients who will work with the information you have created. You might wonder when to credit AI in your work, and if it is even necessary.

Using Agentic AI for work is like using a pocket calculator. It makes you faster and allows you to take on much more complex tasks, but you still need to know the formula to enter, while the calculator handles the calculation and displays the result. I remember my high school teacher telling us, *"Don't rely on your calculator for everything,"* emphasizing the need to retain critical skills like mental arithmetic and to reduce our dependency on technology, for when it was not available or when it was faster not to use it in the first place. AI agents, too, help you and your team process more complex operations and complete tasks faster. You define the goal that the agent should accomplish and provide sufficient context and details about the task at hand. The agent takes over, breaking the goal into subtasks and working on them independently, while you remain in charge of reviewing the result.

	Calculator	**AI Agents**
Purpose	Faster calculation of complex operations	Faster processing of complex operations
Input	Formula	Goal
Skills	Know the exact formula	Frame the goal and constraints
Errors	Incorrect formula or values	Ambiguous instructions or context

Comparison of Deterministic vs. Probabilistic Assistance

A pocket calculator enables its user to calculate the result faster by entering the exact formula and values, while an AI agent accelerates complex processes by interpreting a stated goal and its surrounding context. A calculator depends on precise inputs and clear formulas; any mistakes in the formulas or values lead to incorrect results. Similarly, AI agents depend on well-articulated instructions, constraints, and details. Errors occur when guidance is vague or incomplete. Understanding this distinction clarifies the similarity of what users must provide to ensure accurate results.

As you have seen in Chapter One, using AI is seen as a social stigma in these early days of Agentic AI. Whether you have completed a task independently, with a group of human team members, or with the help of AI agents will become less relevant as AI literacy and maturity increase. In fact, using AI agents will be expected, just as proficiency with Microsoft Office is nowadays.

Shifts in adoption like this do not happen overnight. Instead, they develop over several years and decades. Disclosing when you have used AI, for which tasks, and to what extent is important in the interim to ensure transparency and integrity.

Defining novel concepts and making groundbreaking research discoveries are attributed to individual researchers or research teams. Since the release of AI assistants and agents, the volume of new academic papers has been skyrocketing. That is why universities and publishers of academic journals enforce guidelines that limit or prohibit the use of AI or require authors to disclose the extent of AI use in their research and documentation[30].

In other cases, work products are not directly attributed to individual people but rather to an organization. Corporate blog posts, product documentation, whitepapers, sales contracts, or meeting summaries are good examples of such artifacts. Without a doubt, accuracy is important so recipients act on information that is correct and true, but the author moves into the background as they create information on behalf of the organization. The criteria are different, focusing more on the accuracy and completeness of information for quotes, proposals, or contracts, as well as on consistency and adherence to a common brand voice and style guide for press releases, blog posts, or customer support.

Employees report that their managers overuse AI. When celebrations or holiday wishes are personal in nature, but the

---

[30] Nature, 2023, "Why Nature will not allow the use of generative AI in images and video," June 07, 2023, https://www.nature.com/articles/d41586-023-01546-4.

phrasing, tone, and word choice differ from the manager's usual style, it quickly diminishes the sender's authenticity and trust. Tell your team that you use AI to summarize information that exists in another modality, for example, meeting recordings or transcripts, with links to the actual timestamp in the transcript. That way, anyone receiving the meeting summary can look up the source to get more context. In this example, accurate and objective information is more important than authenticity. AI creates the summary, and you confirm it reflects the discussion and decisions.

Personal notes still require your own tone and style. Access to ChatGPT or Copilot must not become an excuse not to communicate intentionally and authentically—even under the guise of using cutting-edge technology. Not only do recipients expect to hear from *you personally*, but any mismatch in your tone and style (as in the four-sides model) will affect your credibility and authenticity when your teams notice your AI workslop.

Use AI to generate an initial draft or to improve your own. Irrespective of your starting point, you should spend additional time reviewing and editing your draft to ensure it matches your usual tone and style, as you are sending an *authentic signal*. AI is a good starting point for personal communication that requires low authenticity (*executive-ready*) and for institutional communication that requires either high authenticity (*credibility guardrail*) or low authenticity (*fast start*). The difference is the level of human guidance and support in prompting or refining what AI agents generate, so the result is adequate for the target audience.

	Institutional	**Personal**
High	**CREDIBILITY GUARDRAIL** AI-authored, Human-supported	**AUTHENTIC SIGNAL** Human-authored, AI-reviewed
Low	**FAST START** AI-authored, Human-reviewed	**EXECUTIVE-READY** AI-authored, Human-refined

Authenticity (vertical axis) · **Authorship** (horizontal axis)

Types of Agentic AI for Usage

For example, in *executive*-ready scenarios, AI agents create the initial draft, but the human deliberately refines the content to align with their voice, context, and audience. Speed matters more than personal nuance in *fast-start* situations, such as where AI agents generate an initial working draft with minimal human involvement. In *credibility guardrail* scenarios, humans ensure the information is accurate, appropriate, and aligned with expectations to prevent unintended signals that could erode trust if language is misinterpreted. Awareness of when to disclose AI use is a key prerequisite for shaping your AI-ready organization.

Key Takeaways

AI agents challenge the status quo of knowledge work and the business value they create. Leaders actively guide this change.

- Viewing agents as team members is a helpful way to draw parallels to how human teams work along the HUMAN Agentic AI Edge *Operating Model™*.

- Leaders need to act as role models by shifting from effort to purpose, from knowledge to quality, and from praising accomplishments to recognizing impact, to highlight the unique capabilities that human team members possess.

- Professionals typically have a frame of reference about what quality results look like in their domain. AI leads to expectation mismatches between the sender and recipient.

- Depending on the task, context, and audience, disclosing your use of AI is a good practice or simply expected.

- For highly personal messages, use AI as a coach rather than a ghostwriter.

PART II

BUILDING
AI-READY ORGANIZATIONS

Top-down mandates that everyone in the company use AI rarely enable AI readiness, as it depends on more than making the latest technology available to your teams. AI and agents change the requirements for employees' skills, learning structures, business data, and overall governance. As technology expands its scope of information gathering and analysis, the requirements for human expertise evolve as well, shifting toward preparing decision proposals and implementing actions.

Large Language Models (LLMs) consistently improve and excel at answering common questions in professional tests and certifications. This makes them versatile and useful assistants that support a broad range of topics. Nevertheless, humans still need to develop expertise while augmenting their knowledge with AI agents. Digital actions and transactions further improve AI's usefulness in your business. As such, leaders and professionals are codifying their knowledge in the agents they build and maintain.

Slowing entry-level hiring in anticipation of Agentic AI's automation potential is reshaping businesses with unclear short- and long-term effects. Leading hybrid teams of humans and agents presents opportunities to focus on high-value tasks that require human creativity or interpersonal connections. As a sales leader once shared with me, *"Sales is people doing business with people."*

Identifying tasks in which technology augments human capabilities is as important for leaders as assessing the skills of human team members has been to date. Deciding when and how to delegate to agents not only concerns leaders but also professionals. Leaders offer a frame of reference to ensure expectations and results align.

CHAPTER FOUR

CREATING THE FOUNDATION
OF AI-READY TEAMS

For organizations to fully benefit from AI, domain experts across all business functions need to understand AI's capabilities and limitations. This allows stakeholders to identify opportunities for AI technologies based on the status quo they see day in, day out, and on their knowledge of what the technology can accomplish. AI readiness is a journey along four pillars: employees' *skills*, organizational *structure*, business *data*, and *governance*.

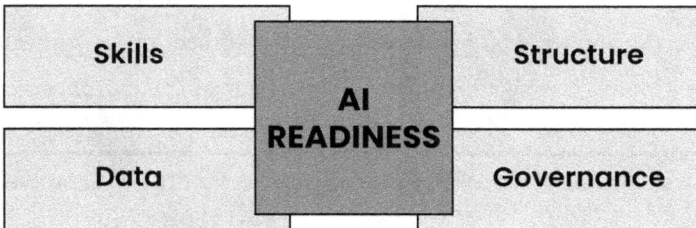

Skills	AI READINESS	Structure
Data		Governance

Core Pillars of AI Readiness

Building Skills That Compound

According to the World Economic Forum, 39% of current skills will be obsolete within the next five years[31], suggesting a rapid transformation of businesses and roles. Many organizations have rolled out ChatGPT, Copilot, or similar AI assistants to their employees in an initial attempt to make AI technology available to their workforce. While these are important productivity tools, simply giving them to employees without proper training and guidelines often results in missed opportunities. Despite the push for companies to become "AI-first," the majority still lack a basic level of AI readiness within their organization. Without this prerequisite, AI adoption happens based on assumptions. Team members "figure out" on their own how to use AI tools. This takes time, happens at various speeds, and with inconsistent results. Additionally, it creates risks of shadow AI, as we have seen earlier.

Leaders are asking which skills are most important to teach as AI agents take on more complex tasks, such as researching information, compiling reports, or completing complex coding projects over extended periods. For example, Anthropic's LLM, Claude Sonnet 4.5, has autonomously worked on a software development project for at least 30 hours straight[32], up from seven

---

[31] World Economic Forum, 2025, "Future of Jobs Report 2025: The jobs of the future – and the skills you need to get them," January 08, 2025, https://www.weforum.org/stories/2025/01/future-of-jobs-report-2025-jobs-of-the-future-and-the-skills-you-need-to-get-them.

[32] Anthropic, 2025, "Introducing Claude Sonnet 4.5," September 29, 2025, https://www.anthropic.com/news/claude-sonnet-4-5.

hours just four months earlier[33]. Naturally, knowledge workers are experiencing the highest degree of disruption and skill irrelevance. Yet the two most critical skills at this point are also the hardest to teach: *adaptability and critical thinking.*

Change is happening at an ever-increasing pace, and stability quickly becomes a burden when it is synonymous with inflexibility. Being resilient in the face of change and being able to adapt are key skills for the future. You will not have all the answers on day one, but you will develop a more complete picture over time. Taking others along on that journey is an important signal to the organization. Leaders build these new skills within their teams by approaching AI literacy along these five dimensions:

1. **Demystification:** Explain what AI is and how it works at a foundational level. Understanding that it is rooted in data, analysis, and prediction reduces common fears that AI will take over the world, which have been shaped over decades by Hollywood science-fiction movies.

2. **Limitations:** AI assistants and agents introduce factual inaccuracies due to the data on which their underlying LLMs are trained, perpetuate biases inherent in their training data, or share data with vendors to improve the technology. Being aware of these risks helps individuals better assess the quality and suitability of the output for a

---

[33] Anthropic, 2025, "Introducing Claude 4," May 22, 2025, https://www.anthropic.com/news/claude-4.

task, whether to refine or revise it, and whether to use a particular AI tool to begin with.

3.   **Context:** Demonstrate how AI agents support common tasks like research, information gathering, and decision-making in the context of business functions, and what the current limitations are. This approach makes AI tangible, and team members see where these capabilities could support their own work.

4.   **Exercise:** Provide individuals with opportunities to learn through hands-on exercises how to prompt AI assistants or build their first agents. When offered in groups, community learning is an added benefit that increases knowledge retention, as more advanced peers share tips and tricks with the group.

5.   **Opportunities:** After learning about the background of current AI technologies and experiencing scenarios for their use, team members are better prepared to determine which tasks and processes within their business function could benefit from AI-driven insight and automation. This context also enables them to become effective partners for technology teams.

Skills compound when the organization learns in layers. Start with three roles that match how work happens along the compounding skill ladder:

1. **Users (baseline literacy)** draft, summarize, research, and spot obvious errors.

2. **Reviewers (quality and risk)** validate claims, check sources, detect bias, and align outputs to standards.

3. **Orchestrators (workflow owners)** break work into tasks, delegate to agents, define acceptance criteria, and run the learning loop that improves performance.

Train users to get output volume, reviewers to build trust, and orchestrators to get scale.

Few organizations become AI-ready through top-down mandates or cascading training along formal hierarchies. Instead, the process is often a lot messier. Some team members experiment with AI tools, developing agents and techniques that work well for them, while others are unaware of the technology or even afraid of it. Companies of all sizes are grappling with this situation as innovation diffusion is a social process, and many leaders are asking how to create an environment that fosters learning.

Creating the Structure for Continuous Growth

As you have seen in Chapter Two, Generative AI and Agentic AI are probabilistic, and identical input (prompts) yields different outputs. Coping with this variance in the results often makes it more challenging for leaders and professionals to quickly and reliably use AI than earlier technologies, and to achieve AI-readiness across the company. There just is no single, golden, or "right" solution to solving a task or problem. Just as in any other business situation, finding a solution depends on communication, context, and iteration—and different approaches lead to the same result. To benefit from the collective experience within their teams and organizations, leaders need to place greater emphasis on sharing information and learning from one another. If one team member is stuck building an AI agent, another team member shares how they approached a similar situation. It reminds me of creating complex Excel formulas and macros in the early 2000s, when everyone had some knowledge and could help each other out. But learning should not be limited to siloed knowledge or knowledge transfer from individual people's minds.

Formal structures give way to more informal ways of organizing information and groups, enabling faster iteration and learning as the core technology, applications, and techniques evolve. Depending on the size of your company, you could organize virtual *lunch-and-learn sessions* held around lunchtime, where a team member shares how they built an agent on a specific topic or how they overcame a particular challenge. This allows other team members to experience the challenges and opportunities in a safe space and to draw inspiration for additional ways they could use Agentic AI. Record the sessions and share the recordings on an internal SharePoint site so those team members who could not join live can watch the replay. It also creates a repository for employees who join your team later to revisit specific concepts or ideas.

In small and medium-sized businesses, *learning groups* are often less formal. As a leader, you personally know the innovative team members by name who are always ready to explore what is next, and you can approach them to help others learn about a new topic like Agentic AI.

In large organizations, professionals with an affinity for technology often start communities within their business function, building informal networks from the bottom up. These *multipliers* organize learning events to share, deepen, and multiply knowledge within the team. Communities like that grow over time and often start in Slack or Microsoft Teams channels. The larger the organization, the more challenging it is to know who is working on what and where, and to coordinate activities. Regional or local subgroups naturally form around innovation experts in these locations. If you do not yet have a forum like this, identify tech-savvy team members and, with their manager's support, ask who is willing to establish a learning community.

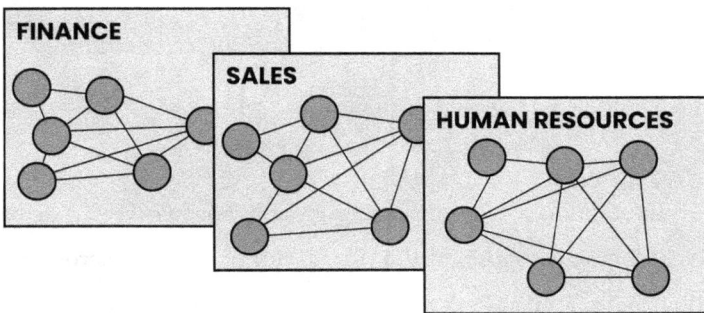

Informal AI Multiplier Groups

If your company is just starting on its AI journey, creating and maintaining a community can be too much effort. Peer learning is an effective alternative. Here is how it works: Pair an early adopter with another team member who is earlier in their learning journey. Ask them to look for one task in their daily workflow that they could automate together with AI and delegate more agency to. As part of this project, both team members learn from each other, enhance their use of AI, and expand the scenarios for which it is used as they learn in tandem. Some organizations run hackathon-like events in which team members identify a business problem in their domain and develop AI agent prototypes to address it within a few hours or a day. This approach works well in teams with a strong competitive spirit, like software development or sales teams.

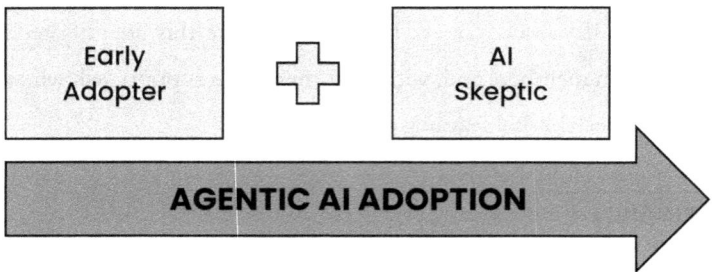

Accelerating Agentic AI Adoption with Peer Learning Groups

In addition to self-organizing groups, hosting weekly or bi-weekly *office hours* provides a forum for colleagues across the company to join and ask questions about available tools and how to troubleshoot common issues. Depending on the size of your company and which department leads the company's AI strategy, the AI lead, head of AI, or a knowledgeable team member could host these regular office hours. These sessions build community as people interact with real people. Set up a recurring meeting invite and publish it on your company portal or your team's SharePoint site. Ask your organization's senior leader to include it in their monthly newsletter to the unit or mention it at a town hall meeting. Forums like these enable team members to ask questions about using AI outside formal hierarchies or impersonal processes. At the same time, those in AI roles learn about the common questions and ideas that business stakeholders have. This opportunity is particularly beneficial to adjust communication plans, end-user documentation, and product decisions.

Turning Business Data into Strategic Advantage

AI labs like OpenAI and Anthropic have been training their models on publicly accessible internet content, including from Twitter and Reddit, and copyrighted material. This has raised concerns about the nature and quality of the content these models have absorbed, including unverified and false information, as well as hate speech, from social media and public internet forums. AI providers have repeatedly shared that they are running out of high-quality, publicly available training data to further improve their models. However, large amounts of the data that AI labs have used to train their models have not even been obtained legally.

High-profile lawsuits, such as a class action against Anthropic for US$1.5 billion[34], The New York Times v. OpenAI[35], or Disney, Universal, and Warner Bros. v. Midjourney[36], underscore that data is a highly valuable commodity. This applies especially if the data has been created with high quality and rigor, and if access to real-time information and events is critical. LLMs have a knowledge cutoff date that limits their usefulness without techniques such as Retrieval-Augmented Generation (RAG).

Several AI labs have been striking licensing deals with content providers. For example, OpenAI has announced deals with The Wall Street Journal, Axel Springer, and News Corp to license their news content. The Walt Disney Company is investing US$1 billion in OpenAI and is licensing its characters to OpenAI for video generation[37]. Google is working with Reddit[38] to use its data. Perplexity has been expanding access to financial filings in the US

---

[34] Reuters, 2025, "US judge preliminarily approves $1.5 billion Anthropic copyright settlement," September 25, 2025, https://www.reuters.com/sustainability/boards-policy-regulation/us-judge-approves-15-billion-anthropic-copyright-settlement-with-authors-2025-09-25.

[35] The New York Times, 2023, "The Times Sues OpenAI and Microsoft Over A.I. Use of Copyrighted Work," December 27, 2023, https://www.nytimes.com/2023/12/27/business/media/new-york-times-open-ai-microsoft-lawsuit.html.

[36] BBC, 2025, "Disney and Universal sue AI firm Midjourney over images," June 11, 2025, https://www.bbc.com/news/articles/cg5vjqdm1ypo.

[37] The New York Times, 2025, "Disney Agrees to Bring Its Characters to OpenAI's Sora Videos," December 11, 2025, https://www.nytimes.com/2025/12/11/business/media/disney-openai-sora-deal.html.

[38] Reuters, 2024, "Exclusive: Reddit in AI content licensing deal with Google," February 22, 2024, https://www.reuters.com/technology/reddit-ai-content-licensing-deal-with-google-sources-say-2024-02-22.

via the Securities and Exchange Commission (SEC)[39] and visual content through a licensing deal with Getty Images[40]. Although this is good news from an intellectual property standpoint, it creates new challenges. Datasets obtained from the public Internet, such as Twitter and Reddit, often contain false information, sarcasm, and potentially harmful content. That becomes a problem when businesses use LLMs for language and knowledge tasks and rely solely on their output. Unlike humans, these models lack the common sense and nuance to distinguish between a sarcastic statement and an actual truth.

One example is the steps that an AI-augmented Google search has recommended in 2024. To prevent cheese from sliding off your pizza, Google recommended adding non-toxic glue. (Do not try this at home. Someone has already done it[41].) The Internet has traced the information back to a decade-old Reddit thread[42] that Google's model has presumably processed and incorporated into its AI-generated output. Additional examples include Google's AI Mode incorrectly conflating information about namesakes, which negatively impacts one because of the other's criminal record.

[39] Perplexity, 2025, "Answers for Every Investor," June 05, 2025, https://www.perplexity.ai/hub/blog/answers-for-every-investor.

[40] Getty Images, 2025, "Getty Images and Perplexity strike multi-year image partnership," October 31, 2025, https://newsroom.gettyimages.com/en/getty-images/getty-images-and-perplexity-strike-multi-year-image-partnership.

[41] Business Insider, 2024, "Google AI said to put glue in pizza — so I made pizza with glue and ate it," May 24, 2024, https://www.businessinsider.com/google-ai-glue-pizza-i-tried-it-2024-5.

[42] Yang, Peter, 2024, "Google AI overview," May 22, 2024, https://x.com/petergyang/status/1793480607198323196.

Luckily, businesses do not need to exclusively rely on the data that LLMs have learned. Companies have petabytes (millions of gigabytes) of data, ranging from master data to transactional data across supply chain, procurement, sales, marketing, and more. Therefore, businesses rarely have a *data* problem, but rather a *sense-making* problem. Because every business is unique, LLMs have limited utility out of the box in individual domains, beyond their excellent language-generating capabilities. In other words, speaking English (or German or French) is not enough when you do not have anything relevant to talk about.

Providing LLMs with data from your company's systems and documents is an option to remediate this problem—but only if you can trust this data to be accurate, current, and complete. Otherwise, you will just end up with more poor information, faster. Garbage in, garbage out. But businesses typically do not have good, readily useful data. In fact, they have had data issues for a long time, spanning data quality, data management, and data governance. As you consider making your own data available to your AI agents, keep in mind that it is likely not as readily usable as you think, and you need to improve it first before it becomes useful.

Software vendors like SAP are pursuing a different route. Based on structured, relational business data, the company has developed a pretrained AI model, RPT-1, that operates on tabular data. The model predicts results for various business problems and objects that previously required significant historical data and individually trained models for each business domain and each company looking to use AI[43]. This approach should further accelerate the use of AI and circumvent the otherwise notorious cold-start problem of insufficient data for model training.

As you are looking to use your business data with AI agents, look for specific, proprietary information about your business that LLMs have not been trained on, such as product documentation, specifications, or pricing. Ensure the dataset is accurate when the AI-generated results are used in your company's production process, in legally binding contracts and policies, or as product metadata. Link AI-generated results to the source for user verification, like the file or URL containing the full details.

Accurate, trusted data is critical for avoiding financial, legal, and reputational damage, whether it is for AI scenarios within your company or for those you make available externally to customers or business partners. AI agents process your company's data to make decisions under uncertainty and evaluate multiple options. Providing them with high-quality data to act on is paramount, as agents make independent decisions significantly faster than someone entering information on a screen, and thereby perpetuate bad decisions across your organization and customers even faster.

Think about autonomous agents that will book your travel, negotiate a contract with your supplier, or provide information

---

[43] SAP, 2025, "A New Paradigm for Enterprise AI: In-Context Learning for Relational Data,", November 04, 2025, https://community.sap.com/t5/technology-blog-posts-by-sap/a-new-paradigm-for-enterprise-ai-in-context-learning-for-relational-data/ba-p/14260221.

about your products, parts, and warranties. Mishaps arising from insufficient data in any of these examples have a real impact on your business, from ending up in the wrong location at the wrong time to overpaying, damaging your customers' assets, and more. That is why spending extra effort to review, clean, and correct your datasets remains key. Similarly, being able to attribute generated information to the exact document or dataset is critical, so your users have a reference point to verify whether the generated output is correct, and you document the source for auditing. Simply adding more data is not the answer. Doing so without focusing on data quality and data governance results in the equivalent business outcome of adding glue to your marinara sauce.

Governing Responsible AI Use

In conversations with CIOs and VPs of IT, they share that their companies already use AI, often developed and implemented by their IT teams. These are real business scenarios in which AI adds value or helps the business team complete tasks they were unable to do before, from supply chain optimization to maintaining product descriptions and automating document processing. But it is not without struggles. Organizational dynamics, politics, and resistance to change are typical barriers to adoption, while bottom-up experimentation with AI and shadow AI weakens enterprise security and supplier negotiation. Larger organizations often create a dedicated AI governance team or committee on which experts from business functions such as IT, legal, security, HR, marketing, and sales contribute and use an intake form to qualify and prioritize incoming requests. In mid-sized companies, the IT team might have an AI governance role in addition to other IT responsibilities.

A simple intake process helps IT and AI teams balance the need for standardization and security with the business's desire to move fast and increase productivity and insights through AI-enabled applications and agents. The key is setting up a lightweight process that takes the requester just a few minutes to complete and the AI team just a few days to review. Employees across the company are more likely to submit information if they expect a quick decision. Establish your intake process by setting up an online form via SurveyMonkey, Google Forms, or Microsoft Forms to capture incoming requests from employees.

Review new requests within the governance team each week and follow up with the requester if any questions arise. Focus on the estimated business value of the solution, potential data, privacy, and security risks, and any similar products you already have a license or subscription for. Approve or reject requests and connect with the requesters of those that were approved one to two quarters after implementing the new AI-enabled product. Validate the initial Return-on-Investment (ROI) assumptions. If the expected ROI has not materialized yet, clarify the root cause and the next steps. For example, give the requester more time to gather additional data points or stop the project and reallocate resources to more impactful initiatives.

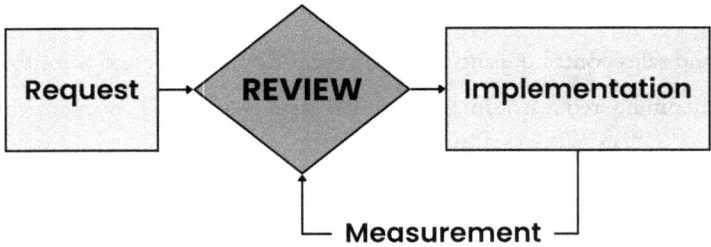

Key Steps in Lightweight Governance Processes

The intake form should include the name and description of the proposed application or agent, as well as the submitter and the business unit or team that will use it. Next, include strategic information such as the project's goal and operational details, including key performance indicators (KPIs) or process performance indicators (PPIs), success metrics, and cost vs. ROI. While KPIs measure the business performance, PPIs focus on the efficiency of a business process. (We will discuss how to measure success, KPIs, and PPIs in Chapter Seven.) Additional information, such as risks, dependencies, data requirements, and the project timeline, provides context for the constraints.

The form concludes with details about the project's key stakeholders, governance, project team, and meeting cadence. (See the sample template on the next page.) The intake or project proposal form is also a valuable resource for reviewing progress and goal achievement relative to the initial assumptions. Use it to assess how close your original estimates were to the actual business impact after the product or feature has been implemented.

AI Project Proposal: *[Project Name]*			Submitter: [Name]
Description			**Business unit** (e.g. Finance, HR, Sales, …)
Strategic goal & benefits	**Business KPI/ PPI**	**Success metrics** (e.g. revenue or savings)	**Cost & ROI** • Subscription • Development • Operations • ROI
Risks	**Dependencies**	**Data requirements**	**Timeline & milestones** • Kick-off • Review • Final delivery
Executive stakeholders • Name, Role • Name, Role • Name, Role	**Governance** • Strategic: Monthly [Date/ Time]	**Project team** • Name, Role • Name, Role • Name, Role	**Operational cadence** • Status review: Weekly [Date/ Time]

Intake Form for AI Project Proposals

Key Takeaways

Organizations do not become AI-ready overnight. It is rather a journey that simultaneously balances skills, structure, data, and governance. In this chapter, we covered:

- Leaders guide their teams to become AI-ready by demystifying and clarifying Agentic AI's limitations and risks, and by contextualizing and providing hands-on exercises to identify AI opportunities in their functions.

- This current wave of AI does not require deep technical knowledge to be useful. Team members organize informal groups or participate in lunch-and-learn sessions, while more formal programs organize learning groups or internal communities.

- Out of the box, LLMs and multi-modal models are good for a broad range of tasks, but it is your company's data about customers, products, and services that makes these models truly valuable in your business. When AI agents use this data to research, prepare proposals, or make recommendations, data quality is important to ensure they provide accurate results.

- Establish a governance process to prioritize internal requests for AI tools, including a digital intake form and a quick review process, that balances agility and risk.

0110 01100001010 001 0111011101011010011010
010101011000000100110011011011001101100111
000111100111010101010010011000010101100101000
011000110000110101011100111011001110110010011
1110000111010111001010100011100110011011011000
11011001001011001000100110010011101 000100
1101100011101111100001011011000000000010111
11101010101 1011011101011010 011 10

CHAPTER FIVE

DEVELOPING
AI-READY LEADERSHIP

Throughout history, existing roles have changed, new ones were created, and others became obsolete because of technological progress. Former switchboard operators, bank tellers, and elevator attendants would certainly have a story to tell. Thanks to Agentic AI, knowledge and insight have never been more broadly accessible, as they are just a few keystrokes away. But when anyone uses these tools, and everyone has access to the same underlying technology and data, human expertise and *how* to use AI become more important differentiators.

Traditionally, humans have instructed software on what to do and what data to use. Now, agents search for information independently, reason, and draw conclusions with limited human involvement, and foundational skills of professions or domains are shifting from humans to agents. As a result, exploring AI agents also needs to include redefining what skills humans need to acquire going forward to become experts and to develop expertise in the first place, and how to best acquire these skills.

A product leader asked whether and how quickly fully autonomous AI agents would be realistic and accepted. The answer depends on increases in both the technology's capabilities (maturity) and users' acceptance of it (trust). Agents currently augment their users' skills through deep research, analysis, and action. As capabilities in reasoning and task duration improve, so will the level of automation. Lastly, once organizations and professionals fully understand the technology's capabilities and limitations, users will be able to determine when to delegate tasks to agents and have them complete them autonomously. This development also evolves the role of humans and human oversight in the process. When AI changes how jobs are done, it also changes the professional identities of those who hold these jobs. Team members are seeking leadership guidance on building meaningful skills and careers as they become orchestrators of work themselves.

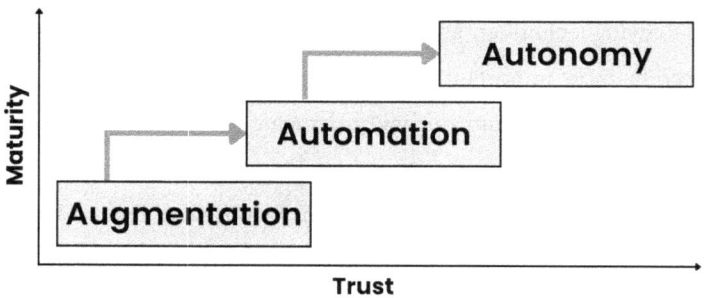

Agentic AI Maturity Levels

Building Experience When AI Accelerates Learning Cycles

In 2025, Dario Amodei, CEO of Anthropic, warned that AI would replace half of all entry-level roles and might lead to a 10–20% increase in unemployment by 2030. Senior leaders outside the technology sector have also adopted the perspective that AI's growing capabilities reduce the number of entry-level roles. However, this thinking assumes that the scope of these roles remains static, while AI agents will take over most of these tasks. Instead, the definition of *entry-level* and experience needs to shift.

Recent graduates will be able to create work products and high-quality results with the help of AI that a professional with a few years of experience delivers today without AI. Likewise, professionals can perform at an expert level without requiring dozens of years of experience, and so on. Like a sliding window, the scope of roles shifts toward combining domain expertise with AI to produce high-quality output.

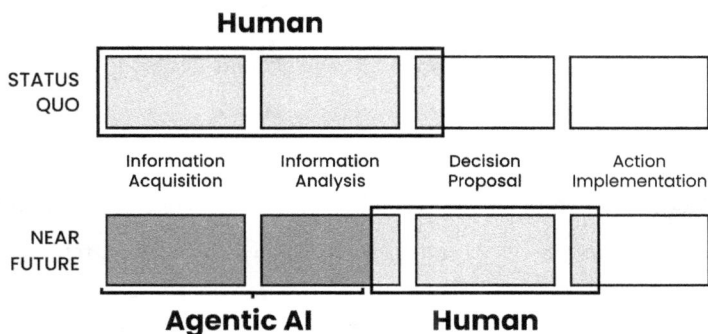

Sliding Window of Human Skills

While roles currently involve acquiring and analyzing information to prepare a decision proposal, Agentic AI completes a large portion of the first two steps, allowing professionals to spend more time on creating and evaluating a proposal, or even taking action within the guardrails of their role.

The first steps toward that future are already happening, as a procurement leader at a multinational semiconductor manufacturer has shared: sending requests for proposals to suppliers and evaluating the bids that come back are crucial steps in sourcing materials and services. Until recently, procurement professionals did most of this work manually. Sourcing materials is time-consuming, especially in large organizations with thousands of suppliers, multiple geographies, and local requirements.

With the help of AI agents, procurement professionals now request this information more autonomously. Human experts define the boundary conditions and goals, and the agents gather information, analyze responses, and recommend actions that the experts review. This approach helps procurement teams scale while focusing their expertise on final reviews and decisions, elevating their role as a strategic partner to the business.

As one of the first companies in the world, life science leader Moderna[44] combined its HR and IT departments under the CHRO in 2025, driven by AI-anticipated change, and started replacing

---

[44] The Wall Street Journal, 2025, "Why Moderna Merged Its Tech and HR Departments," May 12, 2025, https://www.wsj.com/articles/why-moderna-merged-its-tech-and-hr-departments-95318c2a.

junior HR analysts with AI. Even if organizations do not go so far as to combine these business functions, the effect on entry-level roles is already visible across the economy, with fewer roles posted than in earlier years. This is creating a crisis for recent graduates in affected disciplines like HR and computer science.

According to 2025 research by IDC[45] among 5,500 participants globally, the effects of a slowdown in entry-level hiring include increased difficulty in recruiting and training future leaders (71%), widening socioeconomic and workforce diversity gaps (62%), and over-reliance on AI for tasks that benefit from human judgment (48%). Despite the opportunities to reduce costs, the long-term effects on succession planning, on-the-job learning, and the talent pipeline will be significant unless organizations evolve their roles to be augmented by AI rather than replaced by it.

Preparing your workforce for AI use requires more than access to AI tools. Instead, initiatives such as coordinated upskilling programs enable your team members to expand their skill set with AI while building on their existing knowledge and expertise. Without formal training and official top-down support, professionals' skills are at risk of deteriorating. Depending on how individuals use AI, this happens in three possible ways[46]:

---

[45] IDC, 2025, "InfoBrief: AI at Work: The Role of AI in the Global Workforce (commissioned by Deel)," November 2025, https://www.deel.com/the-role-of-ai-in-the-global-workforce.

[46] Abdulnour, Raja-Elie, et al., 2025, "Educational strategies for clinical supervision of artificial intelligence use," August 20, 2025, https://www.nejm.org/doi/full/10.1056/NEJMra2503232.

- **De-skilling:** As professionals increasingly use AI and rely on its output, the actual skill level required to complete a task diminishes. For example, AI-enabled accounting software handles most payment reconciliations. Over time, accountants lose the manual ledger skills to spot errors and complete this task themselves.

- **Mis-skilling:** Professionals blindly trust AI-generated output despite its factual inaccuracies or inherent biases. Rather than applying their own reasoning, over- or under-relying on AI results negatively impacts the overall outcome. Lawyers blindly trusting AI-generated output for their defense without checking whether the sources are valid or even exist is such an example (as we have discussed in Chapter Three).

- **Never-skilling:** Students and professionals just starting their careers might never develop critical skills in their domain that help them reach higher levels of proficiency and expertise. Sales representatives rely on AI-generated prospecting, outreach, presentations, and follow-ups, but they never fully develop the interpersonal skills needed to pitch to clients or understand their needs.

Regardless of roles and seniority, human capabilities such as articulating thought processes, applying critical thinking, and reasoning are crucial when anyone generates an output with AI within seconds. This not only affects early-career individuals.

A common misconception among business leaders is that younger talent teaches older talent, but this is not always true[47]. Early-career talent has a lower barrier to using AI tools, but that does not imply they are experts or have the required domain knowledge to use them effectively. As a leader, you should consider pairing early-career professionals with senior experts to facilitate a knowledge and skill transfer.

The concept of *reverse mentoring* has already proven effective for business skills and extends to AI literacy as well. Following this approach, junior team members gain domain expertise and senior team members benefit from the juniors' experience with AI and agents. Depending on the size of your company, you have a mentoring tool available to support team member pairings. In smaller companies or teams, a project involving a junior and senior team member in your business function can be an effective option.

Before AI's influence on work, the path to acquiring skills was straightforward: you learn a trade, go to college, conduct research, or simply learn by doing. As AI enters the workplace, it takes over more mundane tasks that you and your team members have been learning and doing yourselves. Many professionals wonder where this leaves them. The present is a transitional period in which leaders and organizations need to balance the status quo knowledge with the future skills needed to succeed. Consequently, business

---

[47] Kellogg, Katherine, et al., 2024, "Don't Expect Juniors to Teach Senior Professionals to Use Generative AI: Emerging Technology Risks and Novice AI Risk Mitigation Tactics," June 03, 2024, http://dx.doi.org/10.2139/ssrn.4857373.

and HR leaders need to redefine what it means to be an expert and how to become one. The common definition of expertise is:

$$Expertise = Knowledge + Experience$$

Canadian author Malcolm Gladwell estimated that it takes an average of 10,000 hours (or five years of full-time work) for a novice to become an expert in a field. Others estimate the number at 20,000–25,000 hours (about 10–12 years). Developing from novice to expert level typically happens in five stages[48]. With each stage, individuals' perspectives, decisions, and commitments evolve. From not having a perspective on a subject in the early stages toward becoming experienced, to developing from analytic to intuitive decision-making, and to involvement and commitment, the path shows a familiar progression.

Despite AI's growing capabilities, professionals still need to know the fundamentals of the domain or profession, but they are no longer expected to apply them all by hand. Learning how to work with AI agents is becoming a more in-demand skill than completing the task yourself. Nonetheless, novices still need to gain practical experience to develop expertise and assess AI-generated output that informs their decision-making, *in addition* to achieving productivity gains. AI coding agents like Anthropic Claude Code develop entire applications through prompting, without requiring a software engineering background. This trend, *vibecoding*, illustrates

---

[48] Dreyfus, Stuart, 2004, "The Five-Stage Model of Adult Skill Acquisition," June 2004, https://journals.sagepub.com/doi/10.1177/0270467604264992.

nicely that while anyone can now build applications, the principles of secure software design and efficient programming paradigms remain critical for ensuring that the final application scales and securely handles personal or payment data.

AI changes how experience is formed. If you ignore that, you will create a workforce that generates deliverables quickly but cannot explain, defend, or improve them. Address it by seniority:

- **Early-career professionals:** Require proof of understanding beyond output to build judgment. Ask them to log what they asked the agent, which sources they used, what assumptions they made, and what they verified.

- **Professionals:** Identify the core skills that must stay sharp, such as negotiation logic, customer empathy, domain fundamentals, and risk reasoning. Use agents to accelerate preparation rather than replace thinking.

- **Leaders:** Redesign growth paths, replacing entry-level tasks with supervised "orchestrator apprenticeships" in which early-career professionals run small, agentic workflows with explicit review. This creates experience through responsibility and guardrails.

Although AI shortens the learning curve for novices, they still need thousands of hours of practice to become experts. Productivity increases alone will not develop novices or early talents into experts. Acquiring these skills is possible when the roles and responsibilities involving AI-augmented work are clear.

Orchestrating Human-Agentic AI Work

Managing teams and how work gets done are core responsibilities of any leadership role. As organizations adopt Agentic AI, human team members are now increasingly becoming *orchestrators* of agentic workflows, delegating tasks to teams of agents, coordinating tasks, and monitoring results. Designing agentic workflows and orchestrating work also requires your team to understand when and to what extent to delegate tasks to AI.

Two major patterns to consider are *human-in-the-loop* (HITL) and *human-on-the-loop* (HOTL). In the first one, HITL, a human is part of the process and directly involved in the decision-making or approval of an AI-generated recommendation or proposal. The human approves or adjusts AI-generated outputs. The human has the final say on whether the process will continue and which revisions are required first. In the second option, HOTL, the process essentially runs without human involvement. It is a closed loop in which the human monitors for deviations and intervenes, if necessary, while the AI agent makes the operational decisions.

Human in the Loop

Where do you
need to be ***involved***
to create the result?

Human on the Loop

Where do you
need to be ***aware***
of the result?

Differences Between Orchestration Models

Judging correctly whether a process requires a human *on* (or *in*) the loop significantly reduces your risk exposure. Use *human-in-the-loop* approaches when any of the following are true:

- The outcome changes a person's access, money, employment, or eligibility.
- The output will be shared as a company position.
- The decision is regulated or could create material risk.
- The cost of a mistake exceeds the cost of review.

Design workflows with humans on the loop when the task is reversible, and users detect errors quickly, measure clear thresholds, and review logs periodically. Misjudging when to apply HITL or HOTL leads AI to make a decision that requires human approval or is prohibited by law, such as automatically rejecting a loan application or granting a career promotion.

Regulations such as the EU AI Act place strict limits on the use of automated decision-making in high-risk contexts, requiring appropriate human oversight, governance controls, and the ability for affected individuals to challenge outcomes. Leaders must explicitly design and document where human judgment intervenes, rather than assuming automation alone is acceptable.

Subject matter experts still need subject matter expertise to determine whether the steps an agent takes and the output it generates make sense. Across any role, the critical skills to retain (and develop further) include domain, process, and tool expertise.

Take the process of making price adjustments in retail, where humans provide guardrails and oversight for AI-supported pricing. For example, a small retailer lets an algorithm tweak prices of popular products within strict bounds (±20% of the base price) based on demand or time of day, while a merchandiser monitors dashboards and reviews any unusual price jumps. This human-on-the-loop approach keeps pricing agile and optimized without confusing or alienating customers.

By contrast, Ticketmaster's 2024 handling of Oasis' reunion tour tickets[49] illustrates the effects of using AI-driven dynamic pricing without sufficient human judgment, fairness, and transparency. As millions of fans rushed to buy tickets, the algorithm's surge pricing sent seat prices from about £150 to £355 in real time. With little intervention to moderate these spikes or clearly warn buyers, this approach sparked public outrage and regulatory scrutiny. As a result, Ticketmaster had to rethink its policies and give customers 24 hours' advance notice if dynamic pricing will be used for ticket sales of an event[50].

Orchestrating agentic work is not just limited to the professionals on your team. As a leader, you also delegate tasks to AI agents to support your work. A procurement leader at a leading automotive company shared how they have built a simple briefing

[49] BBC, 2024, "Oasis hit out at Ticketmaster's dynamic pricing after backlash," September 04, 2024, https://www.bbc.com/news/articles/c3w6yy4g6gdo.

[50] BBC, 2025, "Ticketmaster agrees to better price information after Oasis complaints," September 25, 2025, https://www.bbc.com/news/articles/cqxzqvw4lv8o.

agent that compiles information before joining a supplier negotiation that they have not been personally involved in thus far. The agent incorporates information from contracts, proposals, and previous team communications to compile a concise briefing that includes key discussion points and perspectives to anticipate, along with proposals to counter or address them. This agent saves the leader 90–120 minutes of preparation time before each negotiation. While the agent provides a quick overview, the leader reviews the summary's details and verifies the sources the agent analyzed to base its summary on.

This approach allows the leader to spend their time on preparation rather than on information gathering. It is an example of applying the shifting skill window, as discussed in the previous section. Pause and reflect on which of your tasks fit a similar pattern and could benefit from an agentic support.

Orchestration is the new management layer in hybrid teams, encompassing decisions about what to delegate, sequencing agent work, setting review gates, and ensuring accountability. As you orchestrate work and delegate tasks to your agents, review recommendations and decide whether to proceed.

As a leader, you need to be aware of the risks and implications of AI-generated information and assess if AI use is appropriate for the task at hand. You will still need to read and understand the sources to form your opinion, ensure the proposal is feasible, and assess the real-world impact of the decision you are about to take on your business and its stakeholders.

Additionally, you will need to assess the drafts, reports, and proposals your team members send even more critically to ensure the information is accurate, realistic, and feasible.

Managing expectations with your team to deliver high-quality work is a prerequisite, but the risk of workslop creeping into your team's work products is real, and it will fall back onto you as a leader. Therefore, orchestrating agentic workflows requires a close review of the generated information.

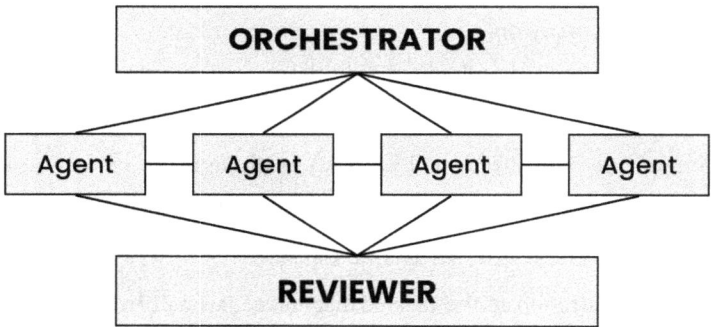

Key Roles in Agentic AI Work

Traditionally, access to information had been siloed. Filing cabinets filled to the top with documents, or limited access to books based on what your local library carried, are good examples of the challenges that predated the digital age. The internet and its worldwide adoption have democratized access to information independent of its location. Now, Generative AI enables anyone to create information faster than ever before, building upon that global connectivity. AI agents further accelerate tasks such as *deep research* by using advanced prompting techniques to structure goals and to self-reflect on interim steps before generating usable results. At the same time, the cost of tasks like these continues to converge toward zero. For example, you can create a hundred variations of a website or draft twelve different versions of a sales email in your CRM application.

The tech industry has solved the *generation of information* at scale. What comes next is the hard part. Filtering, prioritizing, and selecting what is useful is increasingly taking up professionals' time. Evaluating and deciding what hits the mark and what will be discarded propels team members into a reviewer role. The challenge is that humans have traditionally been poor at choosing among dozens of options. Too many choices are daunting and overwhelming, leading to *decision fatigue*. It is the kind of deadlock you find yourself in when you do not know which option to choose quickly. That is why generating more options alone is not helpful.

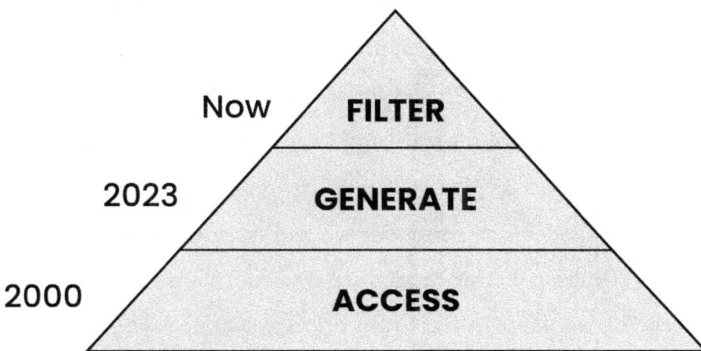

Evolution of Information Processing Challenges

Novel concepts such as *LLM-as-a-Judge* aim to create this filter using AI. The system feeds the output of one AI agent into another that uses a different underlying Large Language Model (LLM). The second agent has specific context and instructions to act as a reviewer and evaluates the first's output. It provides feedback on what to improve and dismisses any low-quality results. This operating model is becoming the new standard in Agentic AI-supported software development. While this is feasible for agents working in a closed loop (until the result is presented to a user), leaders still need to sensitize their team members to be diligent and knowledgeable, regardless of their individual use of AI.

Extending Human Expertise with Agentic AI

Organizations have various kinds of knowledge, and the knowledge captured in information systems often represents only a partial picture of the real world. Institutional and tacit knowledge of how decisions are made, processes run, and customers typically respond exists in employees' minds. Yet it is the daily reality and interactions that shape this experience.

A few years ago, my team and I were leading a product launch. We had created a plan, aligned the timeline, milestones, and messaging, and I was confident we had covered every important aspect. A month before launch day, we met with my manager to review the launch plan and to get his sign-off. He said, *"It's not bad, but it's also not exactly where it needs to be. Can you make it better?"*

We asked for details, but he struggled to articulate exactly what it "should" be and what would make it "better," and responded, *"I can tell you what 'good looks like' when I see it."* So, we went back to the drawing board, wondering what he usually preferred and looked for. After we updated the proposal based on our assumptions, we presented the next iteration. *"It's better than the first one. But can you make it even better?"*

The same cycle repeated itself at least once or twice more, and it was anything but an efficient approach. But the real struggle for me was explaining to my team why, when we presented it to my manager's manager for final approval, they asked why we had not considered the approach that was essentially our initial proposal, before the feedback rounds.

Luckily, with every project, we improved our tacit knowledge and got better at anticipating and predicting what my managers' feedback would likely be. This made the review process more efficient over time.

What seems like an outlier is rather common across companies and industries. For example, a multinational consumer electronics vendor uses AI agents to free up executives' and professionals' time. The company has trained several AI agents based on its executives' typical preferences and feedback and is making them available via self-service. This enables team members to refine their decision proposals before meeting with the executives, accelerating decision-making by 50% and resolving the previous inefficiencies.

In 2025, gig work platform Fiverr introduced a similar concept for freelancers offering their services on the platform. *Fiverr Go*[51] allows freelancers to train an AI model on their work. Customers choosing delivery via Fiverr Go request a logo, infographic, or blog post without depending on the freelancer's actual availability. At the same time, freelancers scale more safely, without AI labs infringing on their copyright. As this work model evolves, so too will full-time roles.

Several startups already enable similar services:

- *Wizly*, a Singapore-based startup, brokers expertise by allowing experts to create their own avatars and train them on their existing content and work products, such as documents, reports, and slide decks[52]. This makes knowledge available globally, independent of the expert's time zone and availability, while the expert gets paid.

- *Personal.ai* offers custom agents based on current knowledge (e.g., your company's documents) and pulls in new data from the Internet as needed[53]. These capabilities are like those of a deep research assistant. One agent excels

---

[51] GlobeNewswire, 2025, "Fiverr Unveils Fiverr Go, a Visionary AI Platform that Puts Creators Front and Center," February 18, 2025, https://www.globenewswire.com/news-release/2025/02/18/3028180/0/en/Fiverr-Unveils-Fiverr-Go-a-Visionary-AI-Platform-that-Puts-Creators-Front-and-Center.html.

[52] Wizly, 2025, "Wizly AI for Independent Consultants and Professionals," December 2025, https://www.wizly.app.

[53] Personal.ai, 2025, "Personal AI—The Distributed Edge AI Platform," December 2025, https://www.personal.ai.

at conducting research, while another excels at applying it to a business function, such as Finance. As a user, add one or both agents to your Microsoft Teams chat and ask them questions. In a future evolution of these concepts, users could add their own content and knowledge and send their avatar to do the work.

As we have seen in Chapter Three, our definitions of work and what it means to work need to evolve and drive further specialization of critical skills and domain expertise. Eventually, professionals will work for more than one company at a time that needs their specialized expertise. Many roles will likely become part-time, requiring specialized industry or domain expertise, while agents handle the repetitive parts. That does not mean there will be two four-hour blocks, with you working in the morning and AI agents in the afternoon, but rather an evolution in how professionals collaborate with AI.

For example, they will complete their work in half the time. Consequently, employers will also pay for just half the workload, like in *fractional* roles or *freelancing*. Income as we know it today, coming from a single employer or occupation for knowledge work, will be the sum of several fractional roles. Together, these income streams will be equivalent to the current income of a full-time role or even higher.

As in previous waves of knowledge management, humans codify their knowledge about the industry, customers, products, and services when using AI agents. In the future, online marketplaces will broker specialized expertise, codified as agents, to complete business tasks and processes. Whether in-house or external agents augment your team, designing and leading Human-Agentic AI teams requires new skills, which we will discuss next.

Key Takeaways

Agentic AI changes how professionals build expertise and how leaders develop an AI-ready organization, from entry-level to experts and orchestrators.

- Like a sliding window, professionals' skills are shifting from gathering and synthesizing information toward preparing decisions and acting on them.
- While AI accelerates tasks, it also presents risks to professionals' and organizations' skill sets, such as de-skilling, mis-skilling, and never-skilling.
- Leaders and professionals are becoming orchestrators of agentic workflows in which humans assign, monitor, and review AI-generated outcomes.
- Based on the risk level involved in a task or process, humans will remain actively involved (in the loop) or take on a reviewer role (on the loop).
- Several platforms and agents make knowledge accessible 24/7, enabling a new economy of fractional expertise.

CHAPTER SIX

LEADING THROUGH AI-DRIVEN CHANGE

S oftware vendors are promising business leaders that AI agents will take over repetitive, mundane tasks and handle more complex tasks. Consequently, many business leaders are looking for ways to reduce spending, transition these tasks to agents, and reduce overall hiring, especially of entry-level roles.

As one CEO shared in a conversation on this trend: *"But I won't hire more people than I need."* That is not the point. But it illustrates the strategic fork that leaders are facing because of AI:

A. Efficiency-only: You reduce headcount and cost and maintain today's product and customer model. This approach risks undermining future capability, succession, and innovation capacity.

B. Capability expansion: You use AI to increase output, then reinvest the reclaimed time into higher-value work, such as deeper customer insight, faster experimentation, better service, and new product motions. This path requires governance and clarity to ensure scalable, consistent, and sustainable impact.

Focusing on expanding your company's capabilities elevates AI from a cost lever into a competitive advantage that enables you to think bigger and to pursue bigger opportunities.

Mid-term, leaders need to consider how team members advance into senior roles, what skills they need, and how to acquire them, given that team members offload parts of their knowledge to AI. That is the kind of mind shift you need to make as a leader of a Human-Agentic AI team. The big question is what you could achieve if the people in your business today had a real productivity booster at their fingertips. Your team members already know your products and services, advise your customers, and live your processes every day. AI agents let them tap into structured research, prepare proposals, weigh options, and run scenario analyses within minutes instead of hours.

Designing Your Organizational Structure for AI Readiness

Companies are organized in hierarchies, with most organizational charts resembling a *pyramid*. Individual contributors at the bottom of the pyramid represent the largest number of employees in the business, followed by first- and mid-level leaders at the levels above. The tip of the pyramid represents the C-suite, including the Chief Executive Officer (CEO).

A slowdown in hiring for entry-level roles (at the bottom) reshapes organizations from a pyramid into a *diamond* and a *kite* shape. Leaders hire fewer junior professionals, and the bottom of the pyramid thins out as current professionals evolve into more senior roles. This approach trades future growth for short-term efficiency gains at a time when most organizations are just beginning to explore Agentic AI and its potential impact on the workforce. Within the next three to five years, cutting back on entry-level role hiring will also pose challenges for succession and workforce planning, as fewer qualified employees are in the business or joining it. These challenges especially apply to large incumbent organizations with several layers of hierarchy.

Leaders often treat workforce fragility as a long-range HR concern. When agents reduce task time and remove visible training moments, gaps in *bench depth, experience pathways, and knowledge capture* appear as sudden quality drift. The team looks productive until the first critical handoff fails, a key person is out, or a customer situation requires judgment that nobody has practiced.

Approach it as you would a customer escalation. Identify the risk, assign a clear owner, set a short recovery timeline, and protect execution time until the issue is stabilized. Then implement a fix that strengthens the process. If the weak indicator is bench depth, build redundancy through shadowing and scenario practice. If it is experience pathways, redesign work so that early-career professionals gain supervised responsibility in bounded workflows. If knowledge capture is the challenge, codify the decision criteria and acceptance standards that determine outcomes, then assign ownership for keeping them current.

Leaders starting a new business augment the expertise of small teams with the scale of hundreds of agents working together, creating a *frontier firm*[54]. In this setup, each team member manages a team of agents to complete projects that previously required a handful of people or more, such as developing software or marketing campaigns. The benefits of this *spear*-like structure lie in scaling through the virtually unlimited capacity of computing resources (agents) and data, at a fraction of the cost of human domain experts. This effectively creates an agentic workforce.

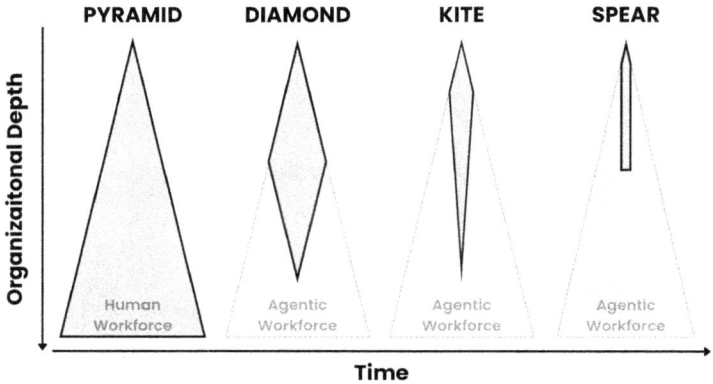

Evolution of Organizational Design

---

[54] Microsoft, 2025, "2025: The year the Frontier Firm is born," April 23, 2025, https://www.microsoft.com/en-us/worklab/work-trend-index/2025-the-year-the-frontier-firm-is-born.

This vision assumes that professionals leading Human-Agentic AI teams have the breadth of industry or product knowledge and domain expertise across multiple specializations. The key skills required for success span defining goals for AI agents, discerning whether the generated results are accurate and appropriate for the task at hand, and combining industry and product knowledge with deep domain expertise. Consequently, the skill profile is evolving from an *I-shaped* profile with deep vertical expertise in one domain to a *T-shaped* profile with additional industry and product knowledge that spans multiple domains. The frontier firm, as well as the business benefits of scaling with Agentic AI, hinge on leaders and professionals developing an *M-shaped* (or comb-shaped) skill profile by deepening expertise across multiple domains.

I-SHAPED PROFILE   T-SHAPED PROFILE   M-SHAPED PROFILE

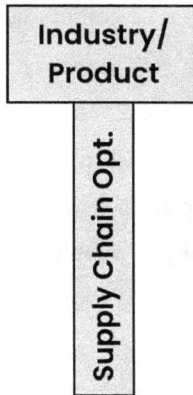

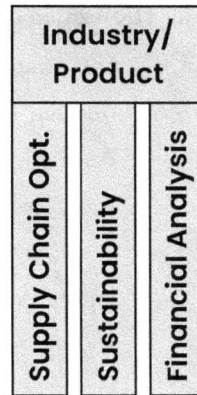

I-SHAPED PROFILE	T-SHAPED PROFILE	M-SHAPED PROFILE
Supply Chain Optimization	Industry/Product — Supply Chain Opt.	Industry/Product — Supply Chain Opt., Sustainability, Financial Analysis

Shift Toward Multidisciplinary Skill Profiles

For example, professionals with an M-shaped skill profile might be highly competent in supply chain optimization, sustainability initiatives, and financial analysis. They make supply chain decisions that are both environmentally responsible and financially sound, while balancing costs and impacts. When overseeing teams of agents, they apply this cross-functional expertise to guide the agents and evaluate and improve their results.

Larger organizations often promote internal talent mobility, in which team members join a different department, such as procurement, engineering, or business development, for a period of time. The professional not only learns about that business function but also brings these learnings back to the department that sent them on this assignment.

Based on your company's size, consider facilitating the following cross-functional learning opportunities together with your HR team in order of time commitment:

- **Job shadowing (1–5 days)** enables a team member to observe and support colleagues in another role or department. The goal is to learn about the typical processes, projects, challenges, and stakeholders.

  For example, a supply chain planner shadows a logistics manager in the warehouse to learn about inbound and outbound logistics processes firsthand, or a software architect shadows a sales rep to learn about typical customer needs and questions, thereby expanding their awareness of customer requirements for new products.

- **Project assignments (1–3 months)** allow team members to contribute to a cross-functional initiative with a defined scope and a limited duration, as part of their actual role. A retail pricing analyst who typically supports the commercial business joins a consumer-division project to learn about customer segments and price sensitivity. This experience expands the analyst's financial modeling skills, which they directly apply in their commercial role.

- **Job rotation (3–6 months)** offers employees the opportunity to move through different roles and departments as part of a structured, time-bound process. For example, apprenticeships (in which apprentices work in different areas of the business) have long been established in trades and, in Europe, are also common in other industries, such as technology. Although it is often a way for entry-level professionals to gain on-the-job experience, you can also adapt the concept for professionals in your company.

- **Job swaps (6–12 months)** let two employees exchange roles for a limited period. Both individuals should have a similar level of seniority, be competent in the domain they are about to enter, and be capable of performing at the level of complexity of the other role.

In a medical device company, a product manager and a technical support lead swap roles, with one resolving clinician escalations and the other handling product backlog decisions. The product manager gains firsthand insight into real-world product issues, while the support lead learns about regulatory and engineering constraints. Both employees can apply the new experience to sharpen prioritization and strengthen customer advocacy in their home roles.

While large companies provide internal talent marketplaces to post and apply for these opportunities, you can also create these learning opportunities in a mid-sized business, together with buy-in from your HR team, senior management, and peer leaders. Established concepts like these let you tap into your workforce's expertise and help team members develop a broader skill set.

Leading Work Across Humans and AI Agents

AI agents not only influence how professionals work. Leaders, too, need to continuously learn, adapt, and inspire others throughout change processes. These skills are more critical than ever as roles evolve and workplace uncertainty increases. The United States National Bureau of Economic Research[55] investigated the success factors for leaders of hybrid teams of

---

[55] Weidmann, Ben, et al., 2025, "Measuring Human Leadership Skills with AI Agents," April 2025, https://www.nber.org/papers/w33662.

humans and AI. Exceptional leaders ask more questions and foster open dialogue with their teams. Additionally, they exhibit strong social intelligence, problem-solving skills, and sound judgment. All these skills have one thing in common: they focus on the human aspects of the organization rather than the technology-driven ones. As the research shows, the leader's ability to navigate the AI transition directly contributes to a team's overall success.

Unless you are in a technology leadership role, deep technical expertise is not a prerequisite. Instead, it is important to stay curious, ask relevant questions, and be open to adjusting how your teams work together. This *learning agility* helps you set an example for your team to follow. But leaders often feel less confident guiding their teams on AI use. Multiple urgent business priorities, tight deadlines, and full calendars are just some of the reasons managers struggle to find time to learn about Agentic AI and gain hands-on experience themselves.

In owner- and founder-led businesses, the top leaders are known for their deep domain expertise and for guiding the company. Any admission that one is less than an expert leads employees to interpret it as a weakness, which diminishes the leader's overall power; at least that is the prevailing view.

Therefore, many leaders rely on news stories, anecdotal evidence, and other secondhand information to guide their views on AI. This approach creates a gap within the company, further exacerbated by poor communication and indecisiveness, or by overly optimistic but empty phrases about the relevance of AI for the business that are detached from reality on the ground, which is little strategic. As a result, employees considering using AI are uncertain whether it is permitted, accepted, or encouraged, and even which tools are safe to use. As a leader, you need to establish clear expectations for AI use, review outputs, and identify cases where human judgment remains critical.

Navigating this environment is quickly becoming an important aspect of being an agile leader. It is okay if you do not have all the answers. However, to be credible when you encourage your team to use AI, you should lead by example and demonstrate it yourself. Start by summarizing meeting minutes with AI and gradually move on to more complex tasks, such as preparing for meetings, conducting competitive research, and analyzing data.

To implement this on your team, dedicate time to experimenting with AI for your daily tasks and to promoting the sharing of experiences. Encourage team members to show how they use AI and explain their problem-solving methods. This fosters curiosity and a culture of learning.

Review your workflows to eliminate unnecessary steps and identify opportunities to increase efficiency, allowing AI to automate more tasks and freeing humans to focus on their core strengths. Keep in mind that AI tools are continually and rapidly improving. Scenarios that are not working today might work within a few months. Adopting a growth mindset will help you and your teams use Agentic AI tools well.

As AI agents become more common in the workplace, it becomes even more important for leaders to adapt to and support others in navigating change. Leading also involves understanding how team members work with AI agents. When your team is new to using these tools, they will have a range of questions.

Build your team's comfort with Agentic AI by addressing the following aspects:

1. **Technology:** When team members are unaware of how agents function or overestimate their agents' abilities, technical problems increase, leading to poor results and AI slop. Provide basic training or collaborate with HR to enhance your team's knowledge and use of Agentic AI.

2. **Processes & Guidelines:** Team members are unsure where or how to implement AI agents into their workflows. Develop a *team AI charter* as a guideline with the team and communicate it clearly and transparently.

3. **Roles & Identity:** AI agents manage complex tasks, such as data collection, synthesis, decision-making, and action-taking. These tasks have traditionally been common in roles that require advanced degrees. When AI systems perform these tasks now, they challenge your team members' professional identities.

   To support your teams, address their concerns and clarify how roles are evolving. The more you invite your team members to shape how their roles evolve together, the more ownership they will feel despite omnipresent uncertainties. This is also why leaders need to increasingly act as coaches.

4. **Expectations:** Senior leaders in your organization also explore Agentic AI and its potential. Depending on your company's culture, they expect significant efficiency gains, although these are not always realistic.

   Be candid about where AI provides advantages and where human supervision is still necessary to shape realistic expectations that you and your teams can meet.

As a leader, you have always motivated your teams to perform at their best and to produce high-quality results. This focus remains regardless of your team's use of AI. Given the rise of Draft Debt, you should reiterate basic assumptions about collaboration and quality that have long been implicitly understood in business. Discuss them with your team, write them down, and periodically review whether the team interacts with each other in line with these values, now encompassing how to work with AI agents.

Start with a simple charter like this, which we will detail in Chapter Eight:

- *We treat AI-generated output as a draft, not as a finished product.*
- *Every team member is responsible for the work they deliver, whether they complete it themselves, with the help of AI, in a group, or with external resources.*

- *We deliver high-quality results that our stakeholders, customers, and partners can depend on without question.*

- *We ask when we are unsure whether AI is appropriate for the task and whether the output is fully understood.*

- *We share our best prompts and agents with each other, so others can learn and benefit as well.*

- *Leaders are responsible for their teams' work.*

Although these aspects apply to all-human teams, adapting them to hybrid teams of humans and AI agents is the next step. You have likely already applied them with your current team. Fostering an inclusive culture where team members share their questions and concerns reduces barriers to responsible AI use.

Human and digital employees operate based on different organizing principles. Understanding these differences is key to introducing AI agents. Treat familiar organizational concepts as design inputs for instructions, constraints, and metrics as part of the HUMAN Agentic AI Edge *Operating Model™.*

1. **Roles** for humans are defined by titles and functions that carry shared meaning. They imply responsibilities, authority, and expectations, often beyond what is written down. AI agents have no such implicit understanding. Their role exists only through explicit instructions and a clearly defined operational scope. Any ambiguity in that scope directly translates into risk.

2. **Knowledge** in human work is built over time through education, experience, and exposure to real-world situations. Much of it is tacit and contextual. AI agents rely on embedded model capabilities, access to specific data sources, and configured reasoning patterns. This knowledge does not evolve organically, and agents do not have an intuitive sense of its limits.

3. **Rules** guide human and agentic behavior, but they function differently. Humans read and interpret policies, procedures, and codes of conduct using judgment; violations are often addressed after the fact. AI agents operate under guardrails and constraints that are enforced at runtime, shifting governance to upfront design.

4. **Rewards** influence behavior in fundamentally different ways. Humans respond to compensation, incentives, and recognition, but imperfectly and unevenly. AI agents respond only to optimization targets, feedback signals, and performance metrics. They will pursue the objective defined, making metric design a leadership responsibility.

5. **Collaboration** among humans relies on team structures, communication norms, and informal coordination. It adapts continuously. AI agents collaborate through explicit interfaces, data exchanges, and workflow orchestration. Every handoff must be designed, as there is no shared context beyond what the system provides.

6. **Organization** ties these elements together. Charts represent human organizations, while informal networks often reflect reality. AI agents require explicit ecosystems that define hierarchy, responsibilities, and boundaries. Without this structure, accountability breaks down.

	Humans	**AI Agents**
Roles	Defined by title and function	Defined by instructions and operational scope
Knowledge	Skills and experience gained over time	Embedded capabilities, data access, and model-based reasoning
Rules	Policies, procedures, codes of conduct, and industry standards	Encoded guardrails, constraints, and operational policies
Rewards	Compensation and performance incentives	Optimization targets, feedback signals, and performance metrics
Collaboration	Team structures, communication norms, and workflows	Agent coordination, system interfaces, and workflow orchestration
Organization	Organizational charts outlining roles and authority	Agent ecosystems specifying hierarchy and responsibilities

Shared Dimensions in the HUMAN Agentic AI Edge Operating Model™

As AI agents become more prevalent, new challenges emerge for leaders and teams. For example, a typical workday consists of tasks with varying degrees of cognitive demand. Some tasks have a higher cognitive load than others. Low-cognitive tasks include copying and pasting data between systems, filing expense reports, and approving routine purchase requests, while designing a go-to-market strategy, negotiating a complex enterprise contract, or diagnosing the root cause of quality issues represent high-cognitive tasks. When AI agents take over most low-cognitive tasks, humans are primarily left with high-cognitive ones.

This approach looks efficient in theory and becomes exhausting in practice, as it leaves fewer mental "breaks" from simpler work. Team members appear productive while decision quality erodes and burnout rises. Every role should include work across three levels of cognitive load:

- **Peak tasks (high):** negotiation, high-stakes decisions, conflict resolution, and strategy, such as leading a pricing negotiation with a strategic customer, resolving a cross-functional delivery dispute, or presenting to executives.

- **Flow tasks (medium):** drafting, structured analysis, and problem-solving with clear boundaries, such as preparing a customer briefing before a quarterly business review or analyzing performance trends to identify improvements.

- **Recovery tasks (low):** organizing materials, closing action items, updating trackers, and administrative follow-through, such as finalizing CRM notes after a sales call or preparing agendas for recurring team meetings.

When agents take over all recovery tasks, teams operate at peak intensity all day. Leaders need to balance automating their team's low- and high-cognitive tasks to optimize performance and results, and reintroduce recovery time through process design.

While every team member now has access to a specialized team of AI agents, using AI reduces collaboration among human team members. As a leader, you should foster meaningful teamwork to avoid isolation. Intentional reviews, face-to-face discussions among team members, or establishing a four-eyes principle process, in which two or more subject matter experts review AI-generated information, create a sense of connection on the team.

AI contributes to new kinds of anxiety and embarrassment when AI-generated meeting summaries misinterpret small talk at the beginning of a meeting as relevant to the actual meeting, leading employees to skip friendly off-topic chats before the meeting begins that would otherwise build community[56]. This inadvertently and negatively influences team culture, requiring your awareness and attention as a leader. In addition to evolving how team members collaborate, understanding when and how to delegate to AI agents is key.

---

[56] Quartz, 2025, "Everyone hates Microsoft Copilot. Does it even matter?," December 07, 2025, https://qz.com/microsoft-copilot-rage.

Deciding What to Delegate and How

AI agents have quickly become the next evolution of automating business tasks and processes, but just like not every business problem is an AI problem, not every task is a perfect fit for delegating to an AI agent. AI agents deliver the best results when tasks involve multiple steps and data from verified sources, and when they produce clear results for humans to review.

Most organizations begin their AI journey with small-scale automation projects, such as customer prospecting and outreach, or market research from public sources. Strategic tasks that impact costs, revenue, or risks are typically good candidates to focus on, especially if you tie them to business benefits that you measure through key performance indicators or process performance indicators, as we have discussed earlier.

Put this into action on your team by creating a list of the tasks and decisions your team spends most of its time on. Prioritize the tasks that have known goals or outcomes, follow defined and repeatable steps, and produce results you can review—and cross out the rest. The remaining items on your list are most likely good candidates for Agentic AI, especially if an AI agent could gather and process data from multiple data sources and recommend or take actions that reduce costs or increase revenue.

For example, supply chain teams manage complex flows of inventory, suppliers, and logistics partners where demand variability, lead times, and risk exposure differ significantly. AI agents connected to enterprise resource planning (ERP), inventory management, and demand forecasting systems analyze historical demand patterns, supplier performance, and real-time inventory levels, identify products or locations at risk of stockouts or excess inventory, quantify the working capital and service-level impact of different replenishment or sourcing decisions, and recommend actions such as adjusting reorder points, rebalancing inventory across locations, prioritizing suppliers, or expediting shipments to stabilize supply and protect revenue.

Good candidates for AI agents typically involve a basic level of complexity, some uncertainty and ambiguity, and benefit from automation. Researching information, organizing data, preparing initial drafts, or identifying prospects to contact follow a pattern, depend on available data, and produce results that are reviewable. However, they are difficult to outline as simple step-by-step instructions.

Humans are essential if tasks involve tacit knowledge and judgment. Setting priorities, weighing trade-offs, interpreting nuance, or making decisions with financial, legal, or reputational impact should remain human-led, even when agents support parts of the work. Humans are also key when agents cannot reliably infer organizational context, relationship history, or ethical implications from the available data alone.

Additionally, consider which tasks your team members generally struggle with due to extensive data analysis or uncovering hidden patterns, and which ones are repetitive yet somewhat complex and do not contribute to overall value.

Effective collaboration with AI means applying many of the principles you are already familiar with from leading people. Clearly articulate the *objective* the agent should achieve, and provide additional *context* and *data* required to complete the task. Additionally, describe any additional agents to *collaborate* with, *deadlines* by which to complete the task, and how you will evaluate the *outcome*. Finally, review and acknowledge the result.

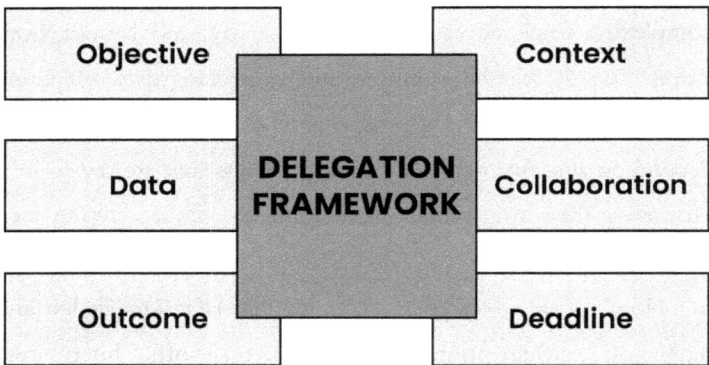

Common Aspects of Task Delegation

Do not forget to also define for yourself what a "good" result looks like, so you do not create Draft Debt. Include these aspects in the instructions you give the agent. And, just as with a human colleague, ask the agent to restate its understanding of the task, the success criteria, and the approach it plans to take.

But it is not just leaders who need to internalize these six steps. Team members who already use AI should apply the same structure to deliver relevant, contextual results. Consider these examples to inspire identifying similar scenarios in your own business or team:

Preparing a quarterly procurement forecast no longer requires manually reconciling supplier spreadsheets, reviewing historical purchase volumes, and assembling a summary for leadership. An AI agent collects spend, contract, and demand data from procurement and ERP systems, analyzes purchasing patterns and supplier performance, and populates a report with updated volume and cost projections. The output serves as a draft for review, refinement, and approval.

Customer service teams also work more efficiently. Instead of manually creating responses, searching knowledge bases, or customizing each reply, an AI agent assesses the incoming request, accesses relevant policies or previous solutions, and drafts a reply. Human team members review and edit this draft to enhance consistency while accelerating their response to the customer.

Staying current on the market and competitive dynamics becomes more efficient in corporate strategy. Instead of manually tracking analyst reports, earnings calls, regulatory updates, and competitor announcements, an agent gathers relevant inputs, synthesizes key signals, and highlights implications for the business. The result is a concise strategic briefing that supports informed decision-making with little manual effort.

Now, expand this to your company:

1. Identify five tasks that your team regularly completes. Focus on those that include information gathering, synthesis, or prioritization.

2. Review the list and determine which of these tasks require adjusting decision criteria as the input changes.

3. From the remaining items, note which tasks involve multiple steps that cannot be fully automated through rigid, predefined rules (*if this happens, then do that*).

4. Assess which of these tasks benefit from automation, drafting proposals, or recommending decisions.

5. Consider which data sources, systems, stakeholders, and decision points influence this task.

6. If all five tasks meet the criteria, consider them candidates for Agentic AI assistance. If only one or a few remain, these represent the most promising starting points for improving outcomes with AI agents.

7. Select one task and define the first action an AI agent should perform.

8. Review AI-generated outputs to verify accuracy, detect gaps, and refine instructions for future iterations.

As you are looking to delegate tasks to an agent, consider whether you would assign this task to a junior colleague, whether you can review the AI's work before sharing it, and if you could explain how the agent generated the result if something failed in the process. Keep in mind that the ultimate responsibility for the final work product remains with the user. That is why you should not delegate tasks that require your personal judgment, experience, or relationships, such as providing feedback, handling customer inquiries, or leading sales discussions. These examples require your human insight and decision-making.

Key Takeaways

Leading through AI-driven change spans skills, roles, and leadership. In this chapter, we learned:

- The organizational design is shifting from a pyramid to a diamond, or even a spear, as fewer entry-level roles enter the business and senior team members are moving up or out of the organization.

- Successful leaders of Human-Agentic AI teams have an M-shaped skill profile that spans deeper knowledge in several domains and broad industry or product knowledge that cuts across them.

- Create cross-functional learning opportunities to help team members build expertise, for example, through job shadowing, project assignments, job rotation, or job swaps.

- Lead your teams of humans and AI agents by developing learning agility that keeps you flexible as technology evolves and changes become more frequent.

- Delegating tasks using a structured approach (from objective to context, data, collaboration, outcome, and deadline) ensures that AI-generated outcomes will meet your needs.
- Focus recurring tasks on AI agents with a basic level of complexity and some ambiguity in their approach.

PART III

SCALING HUMAN-AGENTIC AI AND STRENGTHENING TRUST

L earning about Agentic AI's opportunities for business and designing AI-ready organizations is the foundation for putting it into action across your team or company. Technologists in charge of implementing Agentic AI are defining the future of work as agents take over more complex tasks across business functions. Although many established HR practices and procedures for human employees also apply to AI agents, across governance and risk, talent and workforce management, and rewards and benefits. Yet HR experts are rarely involved in Agentic AI programs, which leads IT and AI teams to reinvent established processes, though with varying levels of consistency and coherence.

For AI investments to deliver a return on investment, leaders and HR teams need to collaborate beyond technology and bring upskilling on AI agents, tools, and best practices to business teams. A key part of these initiatives is fostering responsible AI use without creating AI workslop. Hands-on training provides a common baseline for leaders and professionals to develop simple habits of incorporating AI agents into their daily workflows. Scaling your AI use requires identifying and prioritizing the optimal scenarios that benefit from agents while managing diverse stakeholder expectations. Lastly, noticing and mitigating biases in AI-generated output reinforces responsible and inclusive use of AI and agents. This final section enables you to encourage and empower your teams to use AI and consistently deliver high-quality business results.

CHAPTER SEVEN

EXTENDING PEOPLE SYSTEMS TO AGENTIC AI

H uman resources is an established and highly standardized business function. HR's scope spans determining which business units, departments, roles, and job descriptions a company needs, and which talent, skills, and training support its goals and mission (talent & workforce management), which compensation mechanisms and benefits equip the workforce for success (rewards & benefits), and which policies guide workforce decisions and organizational, regulatory, and operational risk (governance & risk).

TALENT & WORKFORCE MANAGEMENT	REWARDS & BENEFITS

GOVERNANCE & RISK

Core HR Responsibilities

As AI agents complete more complex tasks, act more autonomously, and increasingly operate on your company's behalf, it is helpful to expand these core HR functions to agents as well. Agents change the status quo of how work is done, taking on a single task or collaborating within virtual teams of specialized agents. Introducing agents into a business and its processes changes established paradigms of division of labor, decision-making, and accountability. That is why your company's HR leaders should be directly involved in the transition toward Agentic AI. However, few organizations and HR functions are truly prepared for it.

The *HUMAN Agentic AI Edge Operating Model*™ applies to any entity that creates value in an organization. While many of these aspects have a technical foundation, managing them draws on established organizational and HR principles. If an agent impacts more than one person, treat it like a role that the organization must be able to audit, and document these dimensions in an *agent HR file*.

1. **Roles:** Define an agent's persona, skills, and scope of work. Model an agent as a single task (or role) within a larger system or organizational structure and define the outcomes it is designed to produce.

2. **Knowledge:** Give the agent access to the relevant information for its role and maintain that data source. Monitor performance and user feedback to determine when a knowledge refresh is needed. Capture which data it can use, and which sources are authoritative.

3. **Rules:** Provide the agent with relevant frameworks, regulations, and company policies, such as accounting standards and codes of conduct. Document and spot-check the agent's adherence by reviewing its logs.

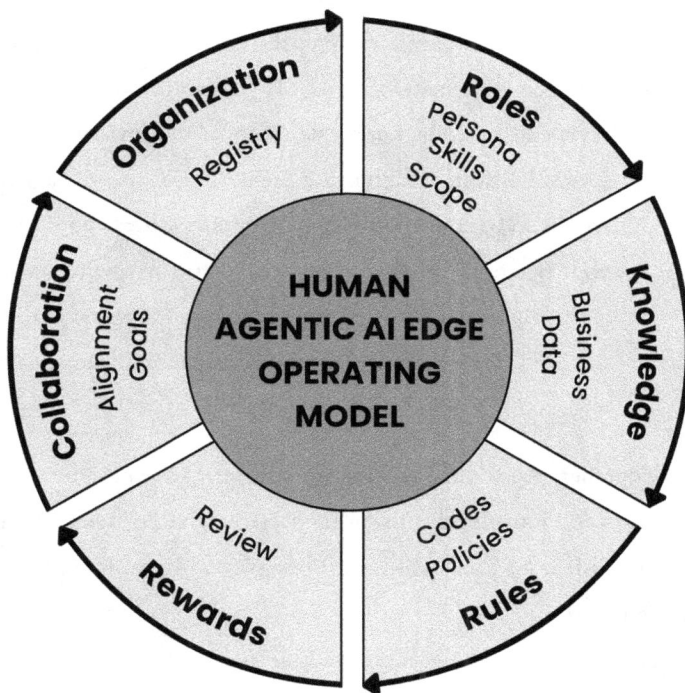

The HUMAN Agentic AI Edge Operating Model™

4. **Rewards:** Regularly measure and monitor the agent's goal attainment to determine whether and how to reward or intervene. Discuss with your technical teams if an agent should receive a raise (higher reward) if it exceeds

expectations, and what to do if an agent does not perform as specified (e.g., update its data or instructions). Monitor performance signals, such as accuracy checks, escalation triggers, and user feedback channels.

5. **Collaboration:** Define goals at the organizational, team, and individual levels to align the agent's reward function with the organization's goals. If an agent underperforms, review its persona, knowledge source, and rewards.

6. **Organization:** Establish a central registry of agents within an organization, like a corporate address book, and your personal subset of the people you frequently work with. This is where agents unfold their full value and, in the long term, their autonomy. Maintain who updates it, who audits it, and how frequently.

Together, these dimensions help you set clear and consistent standards for introducing, using, and governing AI agents—the operations lens of the Dual-Lens Principle we have discussed in Chapter Three.

Although HR experts manage these aspects for humans, they are not the ones adapting them for AI agents. Instead, technologists primarily define what agents do, how they act and react, which personas they adopt, which rules they abide by, and how many agents to create. AI project teams or centers of excellence should include dedicated HR expertise to shape guidelines for your company's AI agents, while focusing on technical aspects.

Setting Standards for Agent Behavior and Risk

Managing agents at scale builds on well-established HR processes for governance and risk. IT and HR teams should collaborate and extend them to AI agents to ensure consistent agent behavior, alignment with corporate policies, and job hierarchies and role definitions.

1. **Consistency of personas and behavior**

    Employers require human employees to abide by a code of conduct to ensure consistent behavior across the workforce. AI agents, too, need to be grounded in a set of common baseline rules as they represent the company and its values.

    When multiple departments independently develop agents, the extent to which they include and enforce such rules varies significantly, leading to inconsistent behavior and results. Define how your digital employees should engage with business partners. The possibility of ensuring this level of consistency is a significant added benefit of AI agents.

    For example, a technology-focused team works with the customer service function to create a team of AI agents. The persona description includes: *"You are a friendly, helpful customer service assistant who does everything to leave the customer satisfied."* When the technology team works with procurement, they define the persona: *"You are an expert in negotiating proposals with suppliers and focus on purchasing the highest-quality products at the lowest price."* How these behave is completely different (making concessions for the customer and ignoring win-win scenarios when negotiating).

There are no central mechanisms yet for making such information available to AI agents and maintaining it. As you develop agents in your business, start with your company's code of conduct, adapt it for AI agents, and store it in a central cloud storage service, such as Microsoft SharePoint, that internal teams and agents can access. Next, ask that anyone building agents reference this code of conduct. Enforce this guideline as part of your governance process for agents used by entire teams or units to ensure consistent, desired behavior.

2. Alignment with corporate policies

Professional codes, licenses, and certifications instill integrity, confidentiality, and objectivity in human employees. Take the accounting guidelines in finance set by the International Financial Reporting Standards (IFRS) or the International Organization for Standardization's standard ISO 20400 for sustainable procurement. As agents take on more tasks, legal and financial risks increase, especially when they operate outside corporate policies and professional codes. Therefore, agents operating in these business functions need to adhere to these policies as well.

For example, an agent's definition broadly refers to your company's code of conduct: *"You should not accept any gifts from customers or prospects or optimize financial gain at the expense of the customer.*" This is just an excerpt of the policy. A hard-coded definition will become outdated the next time the actual policy is updated, and maintaining it across all your agents will require additional effort.

While universally applicable and centrally manageable solutions are still emerging, making corporate and functional policies available via a central cloud storage or collaboration space, such as Microsoft SharePoint, or a central lightweight database (e.g., Airtable) can be an interim solution. Ensure strict version control so that only one approved version of a policy is deployed at any point in time, and that AI agents are grounded in the most recent version.

3. Central job hierarchy and role definition

Companies use organizational charts to define and visualize the formal reporting lines within the organization. Job families, such as marketing strategy or sales operations, group similar roles by function, and job descriptions outline the scope of individual roles, such as senior marketing strategist or vice president of sales operations.

Your company's HR department has job description templates that include standard expectations for individuals in that role, the minimum level of experience and responsibilities, and the other teams or departments they interact with.

A technology team defines an AI agent in simpler and less comprehensive terms, which duplicates the job description:

"You are a customer service agent who provides answers to your customers' most common product questions. You have access to product information from the standard documentation store [DOCUMENTATION] as well as the company's CRM system [CRM]."

Before deploying AI agents across your business, you should determine which role definitions will be centrally defined within a hierarchy based on seniority or role scope. Consider which scope an orchestrator and a reviewer agent need compared to that of a worker agent. Additionally, clarify how to measure and monitor success, and how to reward it.

These aspects become even more important as various teams across your company create virtual teams of AI agents. For example, an *orchestrator agent* must know which agents can answer a particular question or contribute to solving a particular goal. A *reviewer agent* needs specific context and data to analyze, critique, and improve the output generated by a *worker agent*, and a worker agent needs to understand its persona, guardrails, and expectations for the generated output.

These roles and definitions of their capabilities will be similar across business functions. Driving standardization early accelerates the definition of common roles and tasks and ensures consistency across your business. With decades of experience managing *human* resources, the function needs to evolve into a leading advisory function for managing *agentic* resources and shape the future of humans and AI.

As AI agents become the new default in business process automation, they are no longer limited to personal productivity scenarios and rather communicate on behalf of your company. Depending on their persona, an agent helps an employee learn about your company's family leave policy, assists a customer in resolving a product problem, or negotiates an order with your suppliers' agents. Extending governance and risk to agents provides an important baseline for scaling.

Pause for a moment and think about how large volumes of agentic transactions reaching your business at a faster pace could affect it. Consider if your systems and processes are ready to handle the volume of agentic transactions within your organization and from your customers. Modernization efforts, ranging from core business systems like ERP and CRM to evaluating your technology stack and business processes to support an increased volume of agentic transactions, receive renewed focus to help your business thrive rather than be throttled. But futureproofing your business needs to go beyond upgrading systems and requires reconfiguration and rethinking of your business processes.

In addition to the transactional volume, security is another concern. Background checks have become common practice before hiring a new team member to ensure personal safety, verify credentials, and reduce overall risk for the company. Physical security or facility management teams keep detailed records of which employees have received an employee ID card or a physical key that grants them access to a company building. This ensures strict governance and control. Business users connect agents to business systems, repositories, and productivity tools that hold confidential business information. This exposes your company to additional risks when agents are manipulated to take unauthorized actions or simply use a human user's credentials, making them indistinguishable to IT security.

Finally, team members leaving your company become a risk to team productivity if agents in business-critical processes cause unforeseen disruption. That is why any agents that serve more than one user should be well-documented, go through a governance process, and be manageable by your IT team. In addition, new hires need to ramp up and build their own agents.

Managing the Agent Lifecycle

An AI agent's lifecycle is akin to that of a human team member in your company, and it is driven by human-like capabilities (analysis, planning, reasoning, and action) coupled with increasing levels of autonomy and automation. Drawing on HR's *talent and workforce management* practices helps you manage key requirements, processes, and risks as you bring Agentic AI to your teams, preventing the sprawl of an undocumented, agentic shadow organization that extends beyond shifting labor costs through traditional contingent labor models. Eventually, HR and IT will need to collaborate to bring transparency and visibility to this agentic organization, minimizing variability in risk and results. As a leader, draw on familiar concepts from managing humans, ranging from workforce planning to talent acquisition and onboarding and offboarding.

First, define the agent's *persona*, ensuring its communication style and behavior align with your expectations for the role, as you would when assessing a new team member's cultural fit. Then clarify the *scope of work* by outlining the specific tasks and processes the agent is expected to handle, like drafting a job description. Determine the agent's *responsibility level* by setting the boundaries for what it is permitted to decide or execute independently; think of it as its seniority.

Like an employee code of conduct, specify the *policies* the agent must follow to ensure their actions remain compliant and consistent with organizational standards. Describe the *reward mechanisms* that guide its optimization behavior and keep it focused on desired outcomes; compensation is a good equivalent for humans. Conduct strategic *planning* to identify when additional agents are needed based on workload, performance, or business demands, similar to HR's workforce planning for human team members. Finally, outline the organizational structure for agent *collaboration* with teams and other systems, ensuring the right roles, structures, and interactions support reliable, scalable operations.

Several of these aspects are part of HR's traditional scope of workforce management. Applying them to AI agents helps leaders move from ad-hoc experimentation to deliberate system design, with explicit ownership, expectations, and accountability.

	Humans	**AI Agents**
Persona	Culture fit	Role definition and behavior
Scope of Work	Job description	Instructions
Responsibility	Seniority	Permission levels
Policies	Code of conduct	Guardrails
Rewards	Compensation	Optimization targets
Planning	Workforce planning	Capacity planning
Collaboration	Organizational structure	Orchestration model

Transferring Human Management Principles to AI Agents

When any employee creates agents, inefficiencies will increase in how these agents are built and in how they complete common, repeatable tasks. For example, reading, analyzing, and summarizing information from documents is such a repeatable task, whether it is part of a procurement agent that generates RFPs, a market intelligence analyst agent that processes competitors' financial filings, or a sales proposal agent that customizes a contract or offer.

The software engineering discipline addresses this situation by modularizing functionalities that other teams use and reuse. If you are in a technology leadership role, follow this principle for Agentic AI as well. Large organizations set up central agent repositories where reusable worker agents are documented and published for any team member to incorporate into their own agentic workflows and systems.

This approach accelerates agent development and ensures the quality of the underlying agents is both consistent and repeatable; it also adds effort for "non-functional requirements" such as governance, versioning, and coordination, which are often easier to manage in large companies with well-established processes.

Once your AI agents are onboarded and operational, you need to keep them up to date. This is similar to providing training to your employees. Depending on the agents' scope, this means providing them with the latest market data, product or pricing information, regulations, or industry news.

Agents also need to periodically learn about your company's and teams' latest style guides that determine the agents' phrasing, tone, and word choices, as well as policies and professional standards. That means you need to consider when and how frequently to train your agents and to refresh their knowledge repository. While the need to update your agents' knowledge is similar to that of humans, validating that they continue to work as intended afterwards is like a software update; it requires validation and testing before applying it in production.

```
Knowledge  →  Update  →  VALIDATION
    ↑                         
    └──────── Approval ───────┘
```

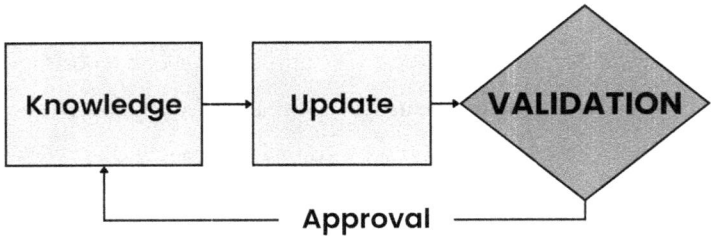

Key Steps in Knowledge Update Processes

Another workforce management aspect is *succession planning*. In addition to the productivity risks of attrition discussed in the previous section, you should define who will take over the agents when an employee leaves.

At a minimum, this serves as an offboarding best practice on your team, and if your company is scaling the use of Agentic AI, it even becomes a formal addition to your company's HR policy and offboarding protocol. This is similar to employees handing over IT equipment and access cards on their last day of employment, and it eventually extends to concepts such as litigation hold and data retention policies to ensure compliance, as well as to reverse-engineering past decisions made by the employee's agents.

Measuring Business Impact

Any investment in AI technology prompts your senior management to request measuring the return on investment. Most software vendors have simplified the cost calculation for their AI assistants sold as Software-as-a-Service subscriptions. As a leader, you can quickly calculate your total subscription cost by multiplying the price a vendor charges (e.g., US$30 per user per month) by the number of users you plan to roll out the AI product to. Estimating the cost for Platform-as-a-Service (PaaS) products typically follows a transaction-based cost model, making it harder to estimate the total cost, which depends on actual usage or consumption (e.g., the number of credits, tokens, or workflow executions).

Immediately after the contract starts, many organizations track the initial investment metrics. *Coverage* indicates the share of employees who will have access to the AI assistant, while *adoption* measures the number of licenses in use relative to the total available licenses. An example of this approach is a study conducted in the UK among 20,000 government employees in the fourth quarter of 2024[57]. Researchers found that users save an average of 26 minutes per day when using AI assistants like Microsoft Copilot, citing an adoption rate of approximately 80% during the experiment. Using metrics like these helps you to demonstrate that you are investing in AI technology and making it available to your teams early in your organization's AI journey. While investment and access are helpful at the beginning, they provide little insight into the business value the AI tools and agents enable.

---

[57] Government Digital Service, 2025, "Microsoft 365 Copilot Experiment: Cross-Government Findings Report," June 02, 2025, https://www.gov.uk/government/publications/microsoft-365-copilot-experiment-cross-government-findings-report.

That is where *usage* data, the next level of indicators, provides additional insights. For example, the volume and frequency with which users with access to the AI product actually use it. For AI assistants like Microsoft Copilot, this could be the number of weekly logins or the number of conversations per user. For AI agents and agentic workflow platforms, the number of completed workflows or processed business objects (e.g., inquiries, applications, or contracts) indicates overall utilization. Additional data, such as total and successful vs. failed executions, provides a more detailed view when your goal is to scale AI usage.

Information such as the *total number of tokens used* helps monitor whether your team is using AI more often or completing more tasks in PaaS scenarios, where incorporating Large Language Models (LLMs) directly into applications is the main goal. These data points are imperfect but valuable proxies for determining whether your teams are using the AI tools and platforms available to them.

Whichever way you choose to capture and report usage data, ensure your process complies with the rules and regulations in the geographies where your users are located (e.g., the General Data Protection Regulation in the European Union). Gathering and reporting anonymized, aggregated usage data could be an option to balance insights and regulations.

Use a third level of metrics to capture the operational *business impact* by focusing on key performance indicators (KPIs) or process performance indicators (PPIs). Take the status quo before using AI as your baseline. Look for industry benchmarks to determine what your company or team should strive to achieve.

For example, finance teams measure a KPI such as Days Sales Outstanding (DSO) to capture the number of days between invoicing a customer and receiving payment. As you introduce AI or AI agents into your business processes, measure the impact on operational efficiency. In another example, measuring the time from service completion to invoice issuance serves as a PPI to monitor and improve. These metrics are leading indicators of success that are immediately measured. However, the *effects* of each category's metrics are lagging indicators, with each one becoming a leading indicator in the next phase.

As with any performance metric, determine which second- and third-order effects arise when adoption and usage metrics trigger AI use that does not add value, or when resources are used on tasks that do not add business value. Top-down AI mandates and encouragement drive some AI use and instill a sense of urgency. But combining them with adequate role- and function-based training and tailoring them to your industry will yield more relevant results. Although the current focus for many organizations is on *productivity*, this is only the first phase. Role-specific AI and agents built into business apps deliver benefits beyond time savings in drafting emails or summarizing meetings.

Phase \ Type	Leading	Lagging
Investment	• Coverage *Licenses vs. employees* • Adoption *Licenses in use vs. total licenses available*	• Usage *e.g., weekly logins, conversations, workflow completions* • Trend *Tokens consumed*
Utilization	• Usage *Weekly logins, conversations, workflow completions* • Savings *Daily time savings* • Trend *Token consumption*	• Process Performance Indicators *e.g., Invoice Cycle Time* • Key Performance Indicators *e.g., Days Sales Outstanding*
Performance	• Process Performance Indicators *e.g., Invoice Cycle Time* • Key Performance Indicators *e.g., Days Sales Outstanding*	• Customer Lifetime Value • Net Promoter Score • Free Cash Flow • Revenue per Employee

Success Metrics and Indicators

Key Takeaways

People systems and HR responsibilities conceptually extend to the agent lifecycle, from governance and risk to talent and workforce management, and rewards and benefits:

- HR has defined processes and procedures for humans— defining roles, knowledge, rules, rewards, collaboration, and organization—that also apply to AI agents.
- Standardize agents and align their behavior with corporate policies to ensure consistency when they act on your company's behalf.
- Define a central job hierarchy that you connect agents to, and job descriptions that anyone building agents in your company can use for their own.
- Agents require clear ownership and accountability, just like human roles, including a named business owner responsible for outcomes and risk.
- AI agents have a lifecycle, including onboarding and offboarding, and leaders need to ensure that agents do not disrupt core business workflows when the employee who built them leaves the team or company.
- People systems provide the structure needed to prevent shadow agents and unmanaged automation from emerging as agentic capabilities scale across teams.

- Initial success metrics for Agentic AI deployments include leading and lagging indicators across the organization's maturity, from investment to utilization and performance.
- HR leaders need to influence and facilitate the shift toward greater agentic autonomy alongside the technologists who are building these capabilities.

CHAPTER EIGHT

MAKING AI A NATURAL
WAY OF WORKING

U sing AI is like riding a bike. You can look at thousands of
pictures of bikes, learn about the mechanics of cogs, chains,
and motion, or watch others ride their bikes. But unless you have
sat on a bike, tried to find your balance, and started pedaling, you
will not be able to move forward, let alone join the Tour de France.
Your team needs to build the habit of using AI to see any
meaningful uptick in usage, return on investment (ROI), and
business value. However, simply telling everyone to use AI "or
else" will not get you there.

As a leader, you have an opportunity to normalize the use of
AI on your team, just like Microsoft Outlook, PowerPoint, and
Word once became the new way of working. Empower and
encourage your team members to ride that metaphorical bike by
giving them proper instructions, training wheels, time to learn and
experiment, and a clear destination—but it starts with you.

Leading Through Change with Clarity

The demands on leaders and their teams continue to increase, and the mantra *"do more with less"* is omnipresent across industries. Yet leaders also need to make time for their own learning and incorporate learning opportunities into their day despite calendars filled with back-to-back meetings. This is challenging, but productivity, insights, and automation do not just apply to individual contributors. It is not realistic for incumbent firms to radically change their operating model from one day to the next and become a frontier firm (as we have seen in Chapter Six).

Many leaders have been promoted to their current roles because of their deep subject-matter expertise and experience. The times when leaders were the top experts in the room are changing, and leaders do not need to have all the answers—they likely never did anyway. Nonetheless, it is getting harder for leaders to hide their own insecurities, which makes embracing AI uncomfortable. It is also only natural that they are concerned their team members will outpace them on the career ladder due to their new AI proficiency. After all, team members can more easily carve out time to learn about AI, especially in organizations with top-down AI mandates and encouragement to increase AI literacy.

Leaders have a special role as change agents within the organization, contextualizing and promoting this change, and communicating that your company wants employees to use AIwithout exposing the company to financial, legal, and reputational risks. Foster this change by modeling the behavior you would like your teams to adopt, setting clear expectations for responsible AI use, and creating feedback loops that demonstrate the desired change and what is possible and practical.

1. **Model the behavior:** Instead of protecting the status quo, leaders need to act as role models and demonstrate the behavior they want their team members to adopt. As many will point out, this is not a particularly new phenomenon, and it applies to change initiatives beyond AI as much as to AI itself. Embracing and leading through change have been core leadership competencies for ages. Leaders need to apply these skills more than ever in this moment and going forward.

   If you do not know where to start, reach out to your HR team to explore organizing a leadership-level introduction to Agentic AI to reduce common reservations and concerns among the leadership team. This also helps increase confidence in using the technology and in identifying where else to apply it. A more detailed adaptation of this hands-on training also enables professionals across the organization to learn more about how to use AI effectively, as we will see in the next section.

   Formally encouraging and empowering your teams to use AI is also important. Signal to your team members that

AI is the new way of working that the organization is moving toward. Build on your own experience with these tools and applications and share how you use them. This helps you build trust and credibility and informs your thinking about where AI adds value within your function.

As a senior executive of a 4,500-person software engineering division shared, it is about leading by example, giving your team members visibility, and recognizing them for their innovative spirit during town halls and meetings. Apply these principles regardless of your team's size.

Gallup research[58] confirms this recommendation. Employees of managers who encourage AI use report twice as much time spent with AI, and they are more than six times more likely to find AI to be useful for their work. These are exactly the kinds of effects that foster responsible AI use in your organization.

2. **Set clear expectations:** Your team members are looking for guidelines for AI use. Develop a simple team AI charter or collaborate with your IT or AI leaders to create a more comprehensive one for your unit or function. Be transparent about when team members should delegate to AI agents, when to seek input from AI, and when to complete tasks independently.

---

[58] Gallup, 2025, "Manager Support Drives Employee AI Adoption," November 08, 2025, https://www.gallup.com/workplace/694682/manager-support-drives-employee-adoption.aspx.

This approach allows you to set expectations that encourage AI use to deliver high-quality results while maintaining accountability for the results. Together, this helps you and your teams be seen as effective and efficient in your work.

3. **Create feedback loops:** Define the boundary conditions for practice, empower your team members to experiment, and lead by example by sharing your own learnings. Team members who experiment with AI share which aspects are working well, motivating others to follow.

Not every idea the team pursues will be suitable for Agentic AI, and not every suitable idea will be successfully implemented or deliver the expected business value.

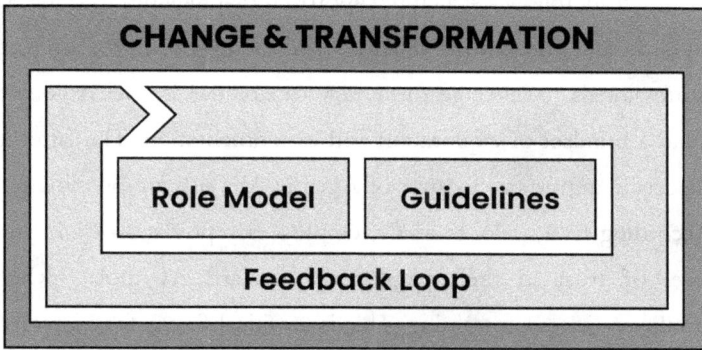

CHANGE & TRANSFORMATION

Role Model | **Guidelines**

Feedback Loop

Leadership Trifecta for Agentic AI Change and Transformation

Upskilling Your Teams to Work Effectively with AI

Organizations are navigating a fine line between promoting AI agents as just another tool in the toolbox and recognizing that using them is fundamentally different from any previous wave of digital technology that leaders and professionals have experienced. Agentic AI opens many more possibilities than applications with a defined feature set have previously enabled.

The absence of the typical constraints of software quickly becomes overwhelming, and professionals ask, *"I don't know what I can use these tools for. Where should I even begin?"* Giving everyone in your company access to ChatGPT or Copilot and leaving them to their own devices to work with AI and agents will not yield the expected productivity boost.

Instead, follow a structured approach that includes baseline AI literacy and hands-on training and workshops to prepare professionals to leverage these new capabilities in their role and build a mindset of exploration and experimentation. The latter is especially important because of Agentic AI's infinite possibilities. According to a Deloitte study, employees report a 144% higher level of trust in their employer's standard AI tools[59] when employers invest in training. This is a critical factor for increased AI use, adoption, and ultimately ROI. But different stakeholder groups in an organization have different training needs.

---

[59] Havard Business Review, 2025, "Workers Don't Trust AI. Here's How Companies Can Change That," November 07, 2025, https://hbr.org/2025/11/workers-dont-trust-ai-heres-how-companies-can-change-that.

- **Stewards:** Your company's *top executives and senior leaders* need to develop a shared vision of AI's strategic impact for the business. While it is easy to agree on the technology's impact on paper, seeing the potential and limitations firsthand guides this group's understanding and decisions on where to invest. It also shapes their credibility as informed AI users and role models across the organization in vendor and employee interactions.

  Initial workshops with this group should aim to clarify the foundations of AI and support alignment among leaders on a shared vision. Next, hands-on exercises introduce participants to prompting, assistants, and agents in the context of the leaders' work. The goal is to show the rapid advancement of the technology and the "art of the possible," while connecting both with the opportunities (and risks) for the business.

- **Orchestrators:** *Mid- and first-level leaders* are the connection between strategy and implementation. They also navigate the tension between communicating and justifying the company's strategic direction to employees, identifying and proving efficiency gains while ensuring high-quality results, and navigating difficult conversations with team members during the change.

  Upskilling programs enable leaders to orchestrate work and workflows across Human-Agentic AI teams. Orchestrators use AI agents for time-saving tasks, such as creating drafts and meeting summaries, and as thought partners to research information, conduct analyses, and explore decision options. The main goal is to increase participants' confidence in using AI and agents and to enable them to support change within their teams by encouraging and empowering them to use AI while remaining accountable for the results.

- **Builders:** Using AI in daily work varies by role. *Software engineers and IT developers* use it to develop new applications and optimize workflows.

  Hands-on training should focus on how to use AI coding assistants and agents available within the concrete software development platforms your company uses. The content blends building AI agents and agentic workflows, and using AI coding tools such as *Cursor, Lovable, Replit, or Microsoft GitHub Copilot X.*

- **Multipliers:** Professionals in this group are known for being at the forefront of exploring the latest technologies within their roles. They are often subject-matter experts who have significant influence within their business functions and help others learn about the opportunities and limitations of AI and agents. Depending on your company's vocabulary, you are familiar with this concept as *early adopters, champions, or power users.* (You can learn how to set up your community of multipliers in the *AI Leadership Handbook.*)

This group needs more detailed examples, current tips and tricks for using your company's AI tools and agents, and strategies for creating high-quality results. Consider periodic updates to keep multipliers engaged and up to date. In larger organizations, multipliers also participate in *train-the-trainer* concepts to scale AI literacy training and enablement in their business functions. Additionally, this group identifies opportunities for AI-augmented process improvements and serves as a feedback channel for the communication and implementation of Agentic AI.

- **Everyday Users:** Most employees fall into this group. They use AI to augment their skills and complete the routine aspects of their day-to-day work more quickly.

Upskilling typically builds a common understanding of AI as a technology and of the specific tools available at your company. Individuals with lower levels of AI literacy are more open to using the technology (as they perceive it as being "magical")[60]. However, demystifying AI agents is a prerequisite for scaling adoption.

Social blended-learning concepts that combine lectures, interactive elements, hands-on exercises, and group work (e.g., in breakout sessions or between modules in a learning journey) enable participants to directly apply what they have learned and strengthen their knowledge.

---

[60] Tully, Stephanie, et al., 2025, "Lower Artificial Intelligence Literacy Predicts Greater AI Receptivity," January 13, 2025, https://doi.org/10.1177/00222429251314491.

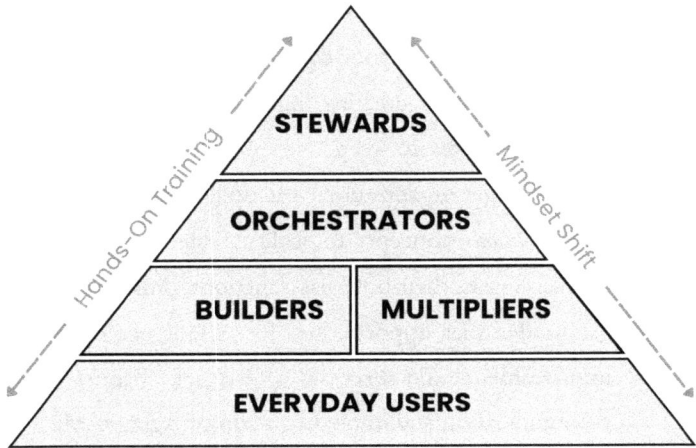

Learner Roles in Agentic AI Upskilling Programs

In addition to the groups themselves, the sequence in which they participate in the upskilling program is critical for cascading knowledge and sponsorship. Top executives and senior leaders need to be aligned on AI's connection and contribution to the company's business strategy. Only with alignment and foundational experience with AI agents and assistants can they be effective stewards in guiding further strategic decisions about AI.

Next in the upskilling rollout plan are mid- and first-level leaders. With their leadership's support, this group effectively orchestrates how work gets done when humans and AI agents collaborate, while ensuring high accountability and high-quality results for their teams. Builders and multipliers receive training in parallel, as they are independent of each other. Finally, trainers or multipliers enable the everyday users.

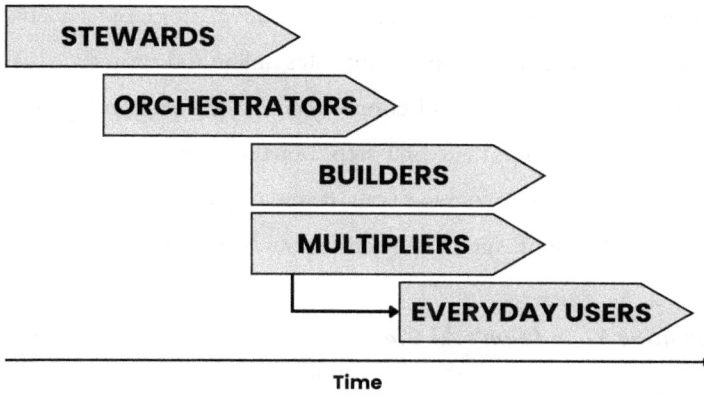

Upskilling Roadmap for Learner Roles

Training must not be a one-time effort or a sporadic benefit. Instead, collaborate with your HR team or an external provider to create a learning journey for each stakeholder group to deepen and broaden their understanding and proficiency at their level. Multinational companies establish formal training programs to standardize learning and upskilling with C-suite support[61]. Based on the number and mix of completed keynotes, training sessions, workshops, and microlearnings, team members reach defined accreditation levels (e.g., bronze, silver, gold) and evolve from *everyday users* to *builders* and *multipliers*. Provide digital certificates and badges to participants upon completing the training, and encourage them to share their learning on internal company portals and social

---

[61] AstraZeneca, 2025, "How we're upskilling and preparing our workforce to thrive in the age of AI," April 11, 2025, https://www.astrazeneca.com/content/astraz/media-centre/articles/2025/upskillingAI.html.

media. Participants of my *Certified AI Leader*™ program frequently share their digital badges and certificates online and appreciate their employer's investment and commitment to learning and growth. Learning is an important first step, but the next is applying it in practice. As you encourage your team to do so, emphasize the responsible use of Agentic AI across your teams or company.

Creating Your Team's AI Charter

For years, the AI community has been debating the fairness and inclusiveness of AI-generated decisions and predictions under terms such as AI ethics, trustworthy AI, and responsible AI. The latter has recently been expanding and also encompasses sustainability, because electricity and water needed to operate the data centers where AI runs impact the environment. Now, we need to expand the definition of responsible AI once again. Not only are developers of AI systems responsible for the effects they have on others, but everyday users are as well when workslop creeps into processes and communications between teams and individuals and causes rework and frustration—the Draft Debt we have discussed.

Shortly after OpenAI released ChatGPT in 2022, one of my team members started exploring it. Michael was learning about prompting techniques and how to get better responses from ChatGPT with fewer iterations. He asked me if he could use the AI assistant as a thought partner to draft multiple versions of a messaging document. Each version would have slightly different angles and tones to cater to different audiences. Michael asked ChatGPT to also provide the pros and cons of each version and to critique itself. At that time, little public information was available on how to write effective prompts. So, he learned most of what he knew through self-study and experimentation. As we were working on AI, using the technology to achieve advanced results more quickly was a logical way to encourage experimenting and learning.

We agreed on two boundary conditions: (1) The output must uniquely match our company's brand voice and the key messages of our product portfolio, and (2) he would share how he arrived at the results in an upcoming team meeting, so others could learn from it as well. Michael agreed, and the results provided a range of options for us to review as a team before discussing our top choice with a professional copywriter.

The team was eager to learn how to develop their prompting skills to support their work, and Michael became the go-to person for effectively using AI tools. Like many leaders, I was focused on strategic deliverables, projects, and the day-to-day business, and could not spend nearly as much time experimenting with this new AI tool myself. But having a curious team member share their learnings accelerated my own learning. Implement this on your team by following these steps:

1. **Encourage team members to use AI** and share their findings and approaches with the rest of the team. Dedicate time in a team meeting for a team member to share. Using AI is *not* a sign of laziness or incompetence.

2. **Set expectations for when, where, and how to use AI**, and tie guidelines, such as your team's AI charter, back to your company's AI policy, including data privacy and approved tools. Go further and discuss with your team when AI supports (e.g., meeting minutes, first drafts) and when humans lead or connect with other humans (e.g., personal contact with customers).

    For example, use AI tools to generate and refine ideas combined with your own research, validation, and expertise. Review AI-generated output like you would review a junior team member's work before passing it on. Passing on AI-generated results without review or fact-checking is unacceptable.

3. **Ensure every team member remains accountable** for the work products they create, whether or not AI has been involved (e.g., reviewing and editing drafts before they go out). Communicating that expectation is key at a time when "good enough" and AI tools are becoming new shortcuts for doing more with less.

    Retain the right to ask for rework. If you spot-check and notice that team members are not following the mutually agreed guidelines, you have the right to ask them to revise. Just as before ChatGPT and similar tools became available, provide feedback and guidance on what should be improved and where the bar is.

Informal practices for using AI form quickly, just like they did with my team member, Michael. Professionals experiment, reuse prompts, and rely on agents in ways that feel efficient individually but diverge when scaled across the team. Without shared guidance, leaders lose visibility into what standards apply and where accountability sits.

Creating a *team AI charter* addresses this gap through a practical agreement that defines how AI is used in daily work. Its purpose is to make expectations explicit and reduce ambiguity through a shared reference for responsible use.

A core function of the charter is clarifying purpose. Team members are seeking guidance that aligns with their workflows, decision types, and risk exposure, as well as general company-level rules. Leaders define why AI is used and where its use should stop.

The charter also establishes shared quality standards and sets clear expectations around accuracy, relevance, clarity, and tone. Additionally, define when team members should disclose their use of AI and how they will review outputs based on risk, ensuring that higher-risk decisions require explicit human oversight. Treat it as a living document to foster your team's responsible use of AI.

Finally, it creates a foundation for discussing what is working well and what needs further improvement, and for adjusting practices as tools and contexts change. Ensure the charter is clear and accessible, for example, on Microsoft SharePoint. It will become second nature as your team's use of AI increases, and it will serve as a key onboarding document for new team members.

Team AI Charter	
Purpose	Why we use AI on this team: *[Insert 1–3 sentences: what we use AI for and what we do not use it for]*
Quality	Every output must meet the following criteria: Accuracy, relevance, usability, and integrity.
Disclosure	We disclose AI use when: *[Insert: external communications; customer-facing content; regulated decisions]*
Data	We never place the following in AI tools: *[Insert: customer data; employee PII; confidential financials; source code]*
Reviews	• **Low risk:** owner verifies before sending. • **Medium risk:** second reviewer validates. • **High risk:** human approval required when: *[Insert: legal/regulatory; customer commitments; pricing; public statements]*
Escalation	If an output fails quality, we: pause, correct, log the lesson, and update prompts/workflows.
Learning loop	Weekly 20-minute review, led by: *[Insert: owner/rotation model]*

Team AI Charter Template

Leaders across any organization play a huge role in scaling AI adoption. Encourage your team to use AI responsibly and take action yourself. For example, start by collaborating with your IT team to identify AI features in applications that your team uses in a process you have identified earlier, and set up a trial to evaluate three to four AI agents.

Identify which systems the AI agent should connect to and what data is required for the task. An agent integrated into your Salesforce CRM accesses customer records and sales data. Similarly, an agent within your document software generates reports or rewrites content. The key is to match each tool to a specific step, ensuring it is used effectively for its purpose.

Invite a few team members to join the trial to gather feedback. Make onboarding simple by asking them to integrate one tool into their daily workflow. Emphasize that the goal is to automate tasks. Participate in the trial yourself with genuine curiosity about evolving work dynamics, especially when team members question changes or are unsure about their future roles.

Lead by example and share how you use AI and agents in your work, and what is working for you, as well as what is not. This approach encourages the team to follow and fosters a culture of shared learning and progress. Expect some initial slowdown as team members get more familiar with AI agents. Investing extra time early pays off later in trust, speed, and quality.

Address quality by setting expectations that AI outputs require review. This helps manage expectations about AI, especially during initial experimentation with agents when team members are not yet aware of the limitations. Establish regular feedback sessions where team members discuss successes and obstacles. Bi-weekly meetings provide a forum to learn from each other. This helps the team become familiar with AI agents and their applications.

Depending on the size of your company and the tool you explore, you could invite a technical expert from IT or your company's AI center of excellence to provide additional support and perspective. As a leader, you should create an environment that welcomes questions about suitable Agentic AI scenarios or about why an agent responded a certain way. This fosters curiosity, skill development, and improved use of AI for appropriate tasks. Once the trial ends, collect feedback to evaluate which Agentic AI tool provides the best ROI and to purchase it for your team.

To encourage responsible AI use on your team:

1. Offer training to lower the barrier of entry—from basic AI literacy to advanced use of tools (as we have seen in the previous section).

2. Empower your team to challenge the status quo of a workflow or process and explore where AI-enabled tools could support them. Adding AI agents to a broken process will not fix it. Instead, it will create more problems faster.

3. Create opportunities for dialogue and exchange by inviting a team member to present at your next team meeting on how they use AI. This provides visibility for innovative ideas and team members and underscores your support that AI is indeed a new skill that team members acquire.

4. Use AI to create drafts that a human reviews, checks, and augments. The value of context and tacit knowledge that human team members add is important here and helps keep them "in the loop" of things.

5. As with any other work product, periodically check AI-generated results to ensure they meet your team's and organization's quality standards. If it does not, provide specific, actionable feedback to the team about what to look for the next time they create similar information.

    High-performing teams only reach that level if expectations and how to exceed them are clear. At the end of the day, your team members remain accountable for their work, whether or not they use AI in the process.

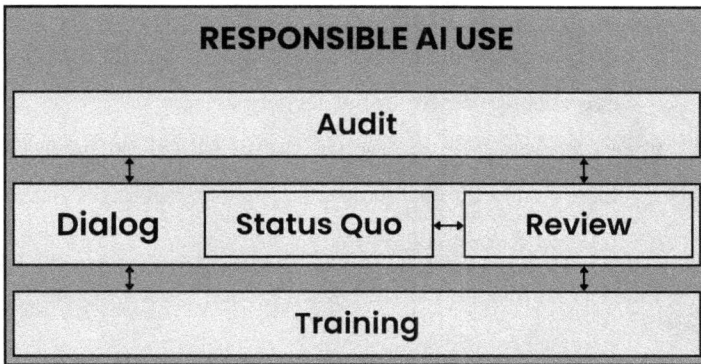

Foundation for Responsible AI Use

Next, we will discuss how to scale your teams' AI use while ensuring high accountability and high-quality results.

Key Takeaways

Leaders act as role models who encourage and empower their teams to use Agentic AI responsibly:

- Leading hybrid teams of humans and AI agents requires curiosity about how this technology works.
- Leaders should invite their team members to share how they use Agentic AI, review which steps in a process can be safely eliminated and automated, and acknowledge that work is changing.
- Provide upskilling for your teams to get hands-on experience with AI agents and understand their challenges and opportunities.
- Training needs vary by stakeholder group, from stewards to orchestrators, builders, multipliers, and everyday users. Top-down awareness and support are critical for adoption and encouragement by subsequent groups.
- Empowering team members to use AI while maintaining high-quality results involves setting expectations about AI use and accountability.

CHAPTER NINE

SCALING AI RESPONSIBLY ACROSS TEAMS

B ringng AI agents to your team creates new opportunities and responsibilities. Agents who work for a single individual or a small team often break when scaled across multiple teams.

Without clear standards, accountability, and shared habits, scale amplifies inconsistency, risk, and Draft Debt instead of productivity. Empowerment quickly degrades into shadow workflows and inconsistent outcomes when technology advances faster than shared standards and organizational readiness. As agents act with greater autonomy, in more complex business processes, and across business functions and heterogeneous systems, biases and errors compound. While agents recommend decisions and act, humans experience the impact of bias and incorrect or fabricated information. Your team can spot individual mistakes made by a single agent, but when multiple agents collaborate on a task, tracing and troubleshooting errors becomes more difficult.

Making Quality a Habit

Imagine the following situation: Your company has provided employees with the latest AI tools and agents, along with a well-defined learning journey to upskill everyone. You model the empowerment and encouragement that support responsible AI use. Yet the monthly reports show stagnation or even a decline in adoption. You start to wonder what is happening. It is like buying an exercise bike to improve your fitness, then never building a routine. The same applies to your use of AI and your team's.

According to social psychology[62], it takes an average of 66 days for humans to form a new habit (with a range of 18 to 254 days). But changing your behavior does not have to be a monumental effort. Small steps (micro habits) help your brain build consistency.

For example, every time you are about to send an email, you add one additional step and ask your AI-enabled writing coach to review it and give feedback. While you need to remind yourself to do it at first, it will become second nature within a few days and weeks. Every time you prepare for a meeting, you ask your agent to summarize the relevant information from across your inbox, chats, and documents. Defining your agent's goals takes some time at first. But once it is set up, preparing for a meeting with the agent becomes just another step in your workflow.

---

[62] Lally, Philippa, et al., 2009, "How are habits formed: Modelling habit formation in the real world," July 16, 2009, https://doi.org/10.1002/ejsp.674.

Scaling Human-Agentic AI collaboration means scaling the habits we have discussed. Over time, habits become behaviors and the default way decisions are made. The combination of *human oversight* and *consequence of error* determines which habits form.

Teams treat AI-generated output as a draft in low-risk situations where humans remain in the loop. Review becomes routine, and mistakes result in rework (*quality issues*) rather than harm. For example, in customer credit reviews, an AI agent summarizes payment history and risk signals, and analysts routinely review and refine the output, with mistakes leading to rework rather than harm. As systems become more autonomous in low-risk settings, such as automatically prioritizing which credit reviews to handle first, habits shift toward speed, allowing inefficiencies and misallocations of effort to accumulate quietly (*efficiency issues*).

The habits that have the most damaging effects form when AI strongly influences high-risk decisions. When the same credit system confidently recommends approving or limiting credit lines, repeated exposure conditions humans to defer to the agents' recommendation rather than challenge it. Over time, analysts stop asking why a limit was suggested and focus instead on processing volume. This over-reliance leads to poor decisions (*judgment risk*) that erode value and trust, such as systematically underpricing risk for customer segments or tightening credit too much in others.

As these risks compound, organizations extend the system's authority, allowing it to automatically adjust credit limits based on real-time behavior. Humans shift from making decisions to monitoring them after the fact. Accountability erodes, and errors surface only once stakeholders are impacted, and complaints accumulate, resulting in legal or reputational damage (*business risk*).

As a leader, scale behavior before you scale value. Governance alone does not compensate for habits that reward speed over scrutiny or delegation over judgment.

Once habits scale, correcting them is far harder than designing them correctly from the start. Pause for a moment and reflect on how you will guide your team members to form quality habits through repeated use of AI.

	Human in the Loop	Human on the Loop
High	**JUDGMENT RISK** Value Loss & Trust Erosion	**BUSINESS RISK** Legal & Reputation
Low	**QUALITY ISSUE** Rework & Inconsistency	**EFFICIENCY ISSUE** Waste & Misallocation

Consequence of Errors (vertical axis, High to Low)

Oversight (horizontal axis)

Decision Matrix for Human Involvement and Risk of Errors

Building your Agentic AI skills is not a one-time effort. To stay effective, you need to apply them regularly, explore new tasks, and adapt as tools change. It is all about building confidence through hands-on experience. Practical tasks help move AI from theory into part of your everyday workflow. The best way to develop your AI skills is to pick one small task in your daily work and explore where AI agents could support you.

Imagine you are a procurement professional evaluating and preparing for a new supplier contract, working with a team of AI agents. Each agent is responsible for a different step in the process—from reviewing performance data and analyzing costs to assessing risks in quality, delivery, and financial stability, and drafting negotiation options. A reviewer agent compares scenarios and highlights trade-offs, acting as your digital support team. However, you must validate insights, adjust assumptions, and finalize the recommendation. Your judgment ensures accurate results and decisions, alignment with corporate policies, and sound supplier choices.

Follow these steps to explore where agents accelerate tasks and where human oversight is necessary:

1. Select one task to delegate to an AI agent.
2. Create the agent and run the task.
3. Closely review the result that the agent produced.
4. Write down what worked and what did not.

5. Confirm that the result is accurate, complete, and aligned with your original goal.

6. Note if and how much you had to edit or change the result.

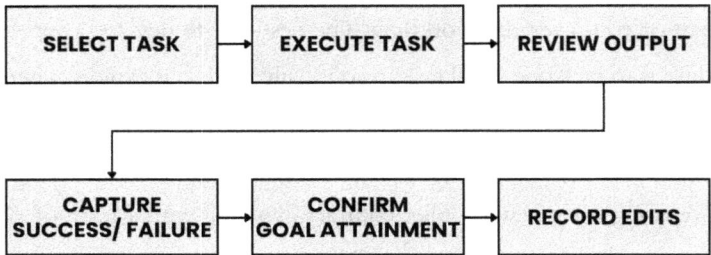

Key Steps for Validating Agentic AI Tasks

Over time, you will notice patterns in tasks that agents perform well and identify where they need improvement. Also, review your notes to decide on your next steps. As you use AI agents more frequently and continue reviewing, adapting, and taking responsibility for the results, agents become trusted parts of your workflow rather than sporadic support tools. AI agents accelerate work, but they do not eliminate the need for human judgment. One of the most important responsibilities you have as a leader is deciding when a human should review the agent's output before it is used, shared, or acted on.

Begin by assessing *what is at stake* in each task. If your team uses the AI-generated output for customer communication, legal documentation, or safety-related decisions, human review is required. Releasing unverified results increases the risk of errors, misinterpretation, or unintended consequences—the results of the Slop Spiral. Once the context is clear, set *defined checkpoints.*

Before any AI-generated content is sent to a client, presented to the board, or published externally, your team should confirm that it is accurate, fair, and appropriate to the context. Also, establish *conditions under which a review may be skipped.* Low-risk internal status updates or early concept sketches do not require immediate review, especially if a more detailed human review occurs later in the process.

Set clear expectations and encourage team members to *flag unusual elements, gaps, or leaps in reasoning* in the agent's response. These signs often indicate that the agent misunderstood the task or used an incorrect data source. Assign *review responsibilities* by role. For example, project managers routinely check agent-generated timelines, while product teams focus on reviewing final feature descriptions before they are shared. Clear guidance tells team members what they are responsible for, when oversight is required, and why it matters. With habits as the foundation, the next step is prioritizing which AI scenarios to explore and pursue.

Choosing Successful Scenarios for AI Use

Leaders are facing lower budgets and fewer resources as pressure on organizations to improve operations increases. In this frequently paradoxical situation, they need to encourage their teams to use AI and create new efficiencies. Reward your team members for experimenting and sharing their results and approaches so others can follow along. Using AI needs to become the new normal. Employees need to know that they can and *should* use AI (in compliance with company policies). Like any change process, this transition will take time, practice, and repetition. That is why your active encouragement and sincere curiosity as a leader are so crucial for shaping that transition.

Your team members are eager to understand how AI agents are changing work, why it matters, and how it will affect them. Guide them to adjust how they work when AI agents become available. Start by providing a simple definition of AI agents as *"a system that can accomplish assigned objectives by autonomously executing required steps. Rather than offering a single response, the agent organizes tasks, retrieves relevant information, and presents outcomes for evaluation."* Next, emphasize that AI agents complement rather than replace human expertise. These systems excel in data analysis, assessing alternatives, and developing recommendations, while interpersonal, organizational, and coordination skills grounded in human agency will remain essential.

View autonomy and decision-making as a continuum. The degree to which an AI agent is involved depends on the role's specific responsibilities. Depending on the role, an agent autonomously manages certain activities, collaborates with human team members, or works under a user's direction to advance the work. For the latter scenarios, the final decision-making, accountability, and judgment remain with the human *using* the agent. Finally, connect this perspective to everyday practices.

As team members become more proficient and confident in using AI agents, delegate additional responsibilities to these systems. Communicate AI agents' capabilities and limitations clearly and establish expectations early with your teams that effective agent use requires learning and adaptation. Remind your team members that collaborating with AI agents is a new way of working for most professionals.

This evolution is an opportunity for you as a leader to go beyond the "AI-first" push and guide your teams to learn together. Your team members who are closest to day-to-day operations will have relevant insights and suggestions to consider when exploring which business scenarios you want to use Agentic AI in.

Despite years of change management, lessons learned, and proven methodologies, just 15–20% of AI projects that you work on will succeed. Agility is key when working on a topic that is moving as fast as AI. What is groundbreaking today will quickly be superseded by faster, cheaper, or higher-performing technology within a few months. Yet despite the rush, it is important to also consider the risks when leading AI projects or programs.

Often, a few key factors along the way separate successful from failed AI initiatives. Few are really technology-induced. For example, one observation has been consistent since 2018. Back then, Gartner estimated that 85% of AI projects do not deliver the value they set out to create. While it is easy to blame a high failure rate on modern technologies like AI, the people affected by the change deserve attention. Team members whose way of working changes due to a new AI capability often oppose the change.

Leaders who have promised quick, successful projects to their superiors are surprised by sudden roadblocks or setbacks in AI projects after status reports have turned from red to orange to green, the higher they rise in the hierarchy. Clear success criteria are missing from the outset. Critical go/no-go decisions are delayed or not made at all, so AI projects continue beyond feasibility, hoping that the next breakthrough is just around the corner—with a bit more time, data, or tweaking. But those are not the only challenges.

One of the most common challenges in any AI project is the assumption that the project flow is linear. It is the typical flow of activities and their connections, whether in school, university, or at work. As a project manager, you divide goals into work packages and estimate start and finish dates, resources, and budgets. The approach applies to researching historical events, printing 3D objects, building an aircraft, constructing skyscrapers, or introducing new software. This also makes any other, non-linear approach seem illogical. (More details on how to overcome the *project mindset* in the *AI Leadership Handbook.*)

But AI projects are, in fact, iterative. Traditional Machine Learning projects have been more akin to research projects when models are built from scratch. Generative AI and Agentic AI projects have lighter data science requirements when engineering teams use the foundation models developed by others.

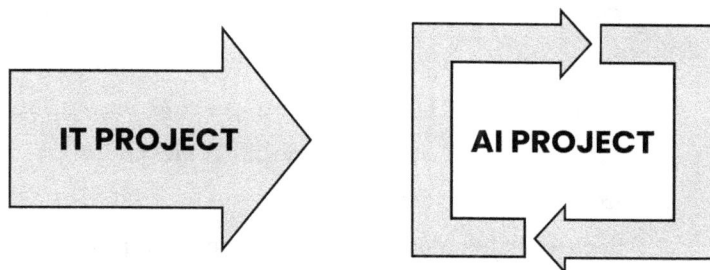

Comparison of Common Project Approaches

Yet ensuring that the AI assistant or agent pulls the correct data, stays within its guardrails, and does not hallucinate still makes the process iterative rather than linear. AI leaders need to manage this expectation with their stakeholders, who are also under pressure to deliver AI-driven results.

These are just some of the common challenges in AI projects. Addressing them upfront requires awareness, foresight, and credibility. The single most important step is clearly defining the business problem and the project's success criteria. Without both, your company will treat AI like a hammer looking for a nail when a different approach is needed (such as using it as a screwdriver).

To overcome the key challenges, follow these steps:

1. **Define the business problem** you are looking to solve. This involves understanding how the solution will help the business stakeholder or project sponsor achieve their goal faster, better, or cheaper, and ultimately help your company succeed.

2. **Quantify business value (and KPIs)** together with subject matter experts of the business function you are working with. This ensures that you can measure the before-and-after states and quantify the improvement from your Agentic AI project.

3. **Secure stakeholder buy-in** early in your project and maintain it throughout, as timelines shift and business priorities change. Review the project status during monthly or quarterly reviews with your business stakeholders and seek formal confirmation that they remain committed to the project.

4. **Assess existing applications** and which vendor's (Agentic) AI capabilities you can use right away to accelerate your time to value. Depending upon your business and industry, the applications you use to run your business's commodity processes in finance, procurement, or HR are good candidates for using *off-the-shelf* AI.

5. **Review data (quality)** to ensure that your AI agents will create relevant results, instead of referencing outdated, incorrect, or incomplete information. This includes operational aspects such as the frequency, test, review, and approval processes to ensure quality and accuracy.

6. **Involve end-users** in your initiatives from the very beginning and throughout the project. Communicate frequently and clearly, and gather their input to enable acceptance and adoption later. Define a vision that your

AI agents support, along with guidance on the expected evolution of tasks, roles, and work.

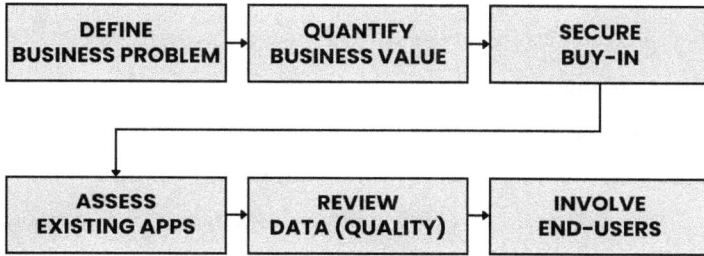

DEFINE BUSINESS PROBLEM	QUANTIFY BUSINESS VALUE	SECURE BUY-IN

ASSESS EXISTING APPS	REVIEW DATA (QUALITY)	INVOLVE END-USERS

Key Steps for Successful AI Projects

In addition to encouraging and empowering team members to use AI and identifying promising scenarios for Agentic AI in your business processes, team members also need to be aware of the limitations and risks of this technology. That is why raising awareness of bias in AI-generated results and how to spot and mitigate it is essential for using AI effectively.

Identifying and Addressing Bias and Errors

Although AI agents seem impressive and handle complex tasks quickly, they also have limitations. The information and recommendations they generate are based on their training data and the instructions they receive. Being aware that agents are neither neutral nor infallible is an important prerequisite for using them responsibly. That is why team members should review AI-generated results with the same attention they would give to the work of a new or more junior colleague, and verify facts, context, and tone before using or sharing any AI-produced content.

Unlike humans, AI tools do not comprehend the real world; they predict what sounds correct based on patterns rather than actual meaning or facts. This results in outdated, irrelevant, incorrect, or biased information, especially when users are unfamiliar with the subject or conditions change quickly.

Biases occur in an agent's output if the training data contains preferences or exclusions. For example, if an agent's underlying Large Language Model (LLM) was trained on Western social media posts, it replicates that tone and vocabulary, even in contexts such as market research, customer service, or job descriptions.

While AI agents support human team members, people remain the ones who share results, make decisions, or approve proposed plans. Just as you would not forward a colleague's draft without reading it first, you should not share an AI's output without verifying it first and understanding what the agent has done.

When summarizing a meeting transcript, AI might generate action items or decisions that highlight the most vocal or assertive speakers, potentially underrepresenting quieter participants. By skewing the focus toward those who spoke more rather than toward those who contributed most, AI excludes important points and divergent views, leading to lower-quality results.

At an AI workshop for manufacturing leaders, I demonstrated how easy it is to find information about physical objects using AI. I uploaded a few pictures of the company's products to ChatGPT and asked it for details about them. While the AI assistant correctly identified the products and their general purpose, the model numbers were incorrect.

As some of the leaders pointed out, ChatGPT referred to models that had been superseded by newer ones several years ago. After updating the prompt and providing a link to the company's product page, ChatGPT correctly summarized the model information. What initially seemed correct and impressive, as far as I could determine as a layperson, was clearly incorrect from the experts' perspectives.

You will likely experience this in other scenarios as well. It will be much easier for you to determine if the generated output is accurate or needs further revision if you already have some knowledge in the domain in which you use AI and agents.

Individuals with limited or no prior domain knowledge, including junior team members or students, lack the same frame of reference as experts and are more likely to accept AI-generated information as adequate and accurate. This presents a challenge if these team members rely too much on AI.

As you saw in the earlier example, double-check the facts. Pause, review, and edit the output if it is too generic, lacks important details or context, is factually incorrect, or does not sound like your authentic voice. Verify that the results align with your team's AI charter and confirm that the potential consequences are clear if a business partner or stakeholder receives AI-generated information that turns out to be incorrect.

AI outputs may contain biases, even if prompts seem inclusive or objective. Regardless of your role, it is important to be aware of this risk and identify when bias could affect your agents' outcomes. Practicing responsible AI use starts with awareness of how these systems operate and where they fall short. Bias emerges in multiple forms, each with distinct implications for decision-making and communication:

- **Representation bias:** Imagine asking an AI agent to summarize global product demand. If the summary disproportionately emphasizes Latin American data while ignoring insights from other regions, it is a clear example of representation bias. This type of bias occurs when the training data is heavily skewed toward a specific group or viewpoint, reducing the diversity of the output.

    To reduce this, expand the range of data sources and refine your prompt to promote balanced results. For example, change your agent's instructions to *"Review insights from all geographic regions and languages."*

- **Stereotyping:** Outputs unintentionally perpetuate social or cultural stereotypes. For example, when using an agent to generate a team bio, it suggests men are *"presentation ninjas,"* and women are *"great communicators,"* revealing embedded biases.

    To reduce such stereotyping, review the output carefully for generalizations related to gender, age, or background, and adjust or remove them as needed.

Providing clear context in your prompt also guides the agent toward producing more neutral and balanced role-based descriptions: *"Ensure the description reflects contributions across different regions, languages, genders, and work context."*

- **Confirmation bias:** At times, the tone or phrasing of AI-generated output suggests favoritism, exclusion, or discrimination. For example, if an agent is asked to create a project brief on switching video editing software, it focuses solely on benefits and ignores trade-offs, leading to biased decisions. This type of bias amplifies existing opinions without presenting alternatives.

  To reduce this bias, instruct the agent to consider both advantages and disadvantages or to present contrasting perspectives and debate them. Update your prompt as follows: *"Draft a brief that covers both the benefits and potential risks of moving to a new video editing application for a mid-sized business."*

- **Omission:** Key perspectives, data points, or facts are completely missing from the output. Imagine using an AI agent to reorder supplies or select equipment before running out of stock. The agent consistently suggests items based on bestsellers or previous purchases, even when they do not fit the current context, or omits better, newer, or cheaper options to choose from. This creates a cycle that reinforces outdated habits and leads to suboptimal outcomes.

To spot this bias, check if the agent has repeated information that does not match your company's or team's guidelines, omitted additional perspectives or facts, or misjudged, misrepresented, or excluded anyone in the generated result.

Apply the Dual-Lens Principle again, this time through the ethics lens. Bias is often subtle and not immediately visible. Therefore, you should implement strategies to reduce bias. Review any AI-generated output thoroughly. Checking the output takes minutes and prevents hours of rework, reputational damage, and exclusionary outcomes when Draft Debt leads to the Slop Spiral:

1. **Source check:** Confirm the source of the underlying data (to the extent possible). This is not always easy, especially if the agent is integrated with your business systems and you do not have access to the underlying data.

   If an agent researches information on your behalf, review the online sources it has accessed to see if the data is public or private, reliable, accurate, and up to date.

2. **Representation check:** Determine who is represented and who is *not*. This could be people, professions, geographic data, or customers. Remember that a confident response is not automatically correct.

3. **Counterfactual check:** Finally, compare the agent's suggestions with your own knowledge and with trusted sources such as known data, benchmarks, or expert input.

If the agent recommends an action, confirm that the reasoning fits your domain, stakeholders, and decision.

4. **Tone check:** Ensure language is appropriate for the audience, culture, and context.

5. **Decision ownership:** Document who approved the output and why it was acceptable.

In case the AI-generated result contains bias, adjust the agent's goal. Instead of asking, *"What are the latest retail pricing trends?"* say, *"What pricing approaches do big box retailers use for private label clothing lines?"* This leads to a narrower, more specific exploration.

You cannot detect and remediate all bias, but you should try to recognize its presence and strive to mitigate it. If editing the results becomes too time-consuming, start over or conclude that this task is not well-suited for an AI agent (yet).

Especially for sensitive topics, personal judgments, or confidential information, ensure more human oversight or address them manually yourself. If in doubt, review the guidelines available on your company's internal portal and ask your manager.

In the final chapter, we will look at where Agentic AI is headed, how it evolves, and how leaders continue to shape Human-Agentic AI teams in this transformation.

Key Takeaways

Scaling Agentic AI from a handful of ideas to a new way of working across AI-ready teams takes effort:

- Hone your AI skills and improve your approach with every iteration to discover additional opportunities.

- Put AI into action by setting expectations with your team in a team AI charter that addresses the shared values, tasks, empathy, and ethics.

- Communicate the guidelines clearly and share them with your team. Emphasize that every team member is responsible for the results they create; reviewing them before passing them on to others is a critical step.

- Introducing AI into your team is unlike any other project that you have previously been involved in. AI projects are iterative in nature (as opposed to a linear flow).

- Underestimating complexity and stakeholder buy-in are among the key reasons why 85% of AI projects still fail.

- AI Agents and the underlying LLMs introduce bias into their outputs. Sensitize your team members to this situation and to how to notice and correct it. That way, the information is more representative and inclusive of reality.

PREPARING THE HUMAN-AGENTIC AI FUTURE

A I innovation is developing rapidly. Keep your knowledge current to use Agentic AI responsibly and effectively in your role. What seems impossible today might soon become achievable, and current capabilities could become new standards. It is helpful to find easy ways to stay informed without feeling overwhelmed.

A key trend to watch is the growing autonomy of AI agents, taking actions across different tools, systems, and data sources. They will be able to handle more complex tasks within set boundaries. For example, in finance, agents reconcile transactions, spot anomalies, and prepare reports with little human help. That is why understanding how much control you want to give and how much you prefer to keep under your own supervision is so important for adjusting your strategy based on your risk level and intended target audience.

Expanding Agentic AI Across Organizational Boundaries

AI agents are also becoming more collaborative. From single functional tasks to collaboration within and between departments, teams of specialized agents are completing complex processes with minimal human oversight. For example, in customer service, an orchestrator agent analyzes an inquiry and routes it to the right support agent for processing. This worker agent gathers product info and drafts a response, and a reviewer agent checks the draft for accuracy and tone. This teamwork among agents makes handling advanced interactions faster and more seamless, while humans remain in control for final checks and sign-off.

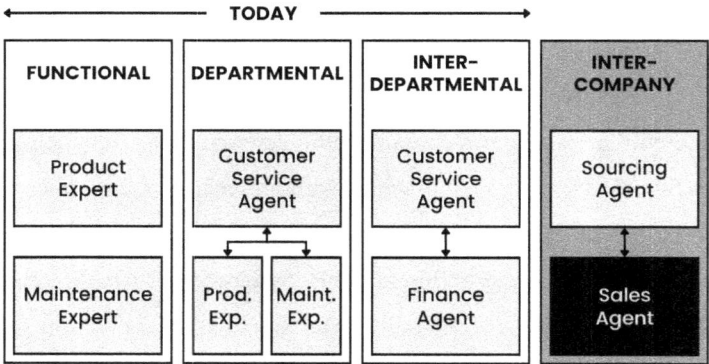

Scope Increase of Agentic AI Across Organizational Boundaries

In the future, agents will also complete business transactions on behalf of their companies, such as a sales and a sourcing agent negotiating terms and replenishment orders. The concepts we have discussed in this book will be even more relevant as agents' autonomy increases. That is why establishing governance early in your company's Agentic AI journey sets the foundation for future scale. As you develop and use AI agents, carefully manage the permissions you grant them. The goal is to make sure they operate within clear limits and that vendors follow strict security and compliance rules. It is important to understand how your data is stored, processed, and protected before letting agents work freely in sensitive parts of your workflow.

Many organizations are putting in place internal guidelines for responsible AI use. These include being transparent, documenting where and how AI is used, and setting rules on sharing information outside the organization. Following these policies builds trust and accountability.

Also, staying informed through applied AI newsletters, setting monthly reminders to test new features, attending webinars or training sessions, and connecting with peers or communities that share experiences are helpful.

Understanding upcoming roadmap updates and their relevance to your role helps you stay ahead. Taking the time to learn about the latest trends enables you to contribute meaningfully and adapt smoothly as AI tools and agents continue to evolve quickly. However, you will need to define a filter to separate the signal from the noise, for example, by focusing on information relevant to your industry, business function, and geography. Keep in mind that not every headline requires immediate attention.

Designing Human-Centered AI Experiences

As we discussed in previous chapters, Agentic AI presents a fundamental shift in how work is done. This applies not only to the paradigm shift in researching and executing tasks but also to the user experience of carrying them out.

Let's compare how humans work together to see how it applies to agents. Jane, in accounting, has access to several applications and IT systems, including Microsoft Office, ERP, SharePoint sites, and analytics dashboards. You reach out to Jane via email to ask whether your suppliers' invoices have already been paid. She has the language and domain skills to understand your request and the access to the tools to provide the answer. Instead of pointing you to one ERP system for supplier A and another for supplier B, or responding via email for one and via Teams for the other, she sends you the details in a single email reply.

Agentic experiences still feel fragmented today, but they will evolve over time. Agents and their user experience need to become interoperable for them to be even more valuable to business users. Software vendors like Microsoft, Salesforce, and SAP are embedding AI agents directly into apps you already use, including productivity tools and ERP and CRM systems. These embedded agents help with scheduling, drafting, analysis, and task completion, without requiring you to switch between applications, while making AI more accessible. These AI features are part of the operational excellence category we discussed earlier.

Where you start your inquiry or task is less relevant. If a user starts a task in Microsoft Copilot and it needs access to financial information in an ERP system from a different vendor (e.g., SAP or Oracle), it queries that system to retrieve the necessary data, providing a seamless user experience. Vendors are laying the groundwork through open protocols and standards[63].

Beyond task-level user experience, the second evolution is already underway, beginning with consumers. Users are turning to AI assistants and agents for web search and deep research insights, and companies with applications that have hundreds of millions of monthly active users, like ChatGPT, are becoming the new user experience layer.

AI labs' partnerships with omni-channel retailers like Walmart[64] and payment platforms like PayPal[65], in conjunction with agentic browsers, spotlight e-commerce as the next frontier for Agentic AI. Browsers like OpenAI ChatGPT Atlas or Perplexity Comet perform tasks on a user's behalf, including navigating websites. While this new user experience begins with common consumer transactions such as e-commerce and travel

---

[63] Microsoft, 2025, "Adobe, SAP, and ServiceNow talk agents and Microsoft 365 Co-pilot," May 20, 2025, https://www.youtu.be/0wbxKei7EI0.

[64] Walmart, 2025, "Walmart Partners with OpenAI to Create AI-First Shopping Experiences," October 14, 2025, https://corporate.walmart.com/news/2025/10/14/walmart-partners-with-openai-to-create-ai-first-shopping-experiences.

[65] PayPal, 2025, "OpenAI and PayPal Team Up to Power Instant Checkout and Agentic Commerce in ChatGPT," October 28, 2025, https://newsroom.paypal-corp.com/2025-10-28-OpenAI-and-PayPal-Team-Up-to-Power-Instant-Checkout-and-Agentic-Commerce-in-ChatGPT.

booking, businesses will soon automate equivalent decisions and processes that previously required manual steps and systems.

AI labs have added web search to their AI assistants to provide current data beyond what their Large Language Models (LLMs) have been trained on and to generate more relevant results. Perplexity has built its company around this value proposition, and Google rolled out its AI Mode, which summarizes information about a search topic.

As a result, digital marketing teams and agencies have been adapting to this shift, as Search Engine Optimization (SEO) techniques and tactics developed since the early 2000s have become less relevant. Topic authority and relevance are changing again, moving from crawlers and bots to AI assistants.

Similarly, your company's marketing department needs to evolve its practices as AI assistants and agents continue to recognize your company's authority. As agentic browsers become the new gateway for e-commerce transactions, your website's layout, design, navigation, and operation need to evolve just as much. You no longer design for humans. Agents prioritize based on functionality and structure.

Companies like Amazon have taken a critical eye to this vision for e-commerce and have taken legal action against Perplexity over its agentic browser, Comet. Amazon fears disintermediation of its ad business if agents, rather than human customers, visit its website or use its app for shopping. Fewer humans using these interfaces means displaying and acting on fewer ads, which will shrink Amazon's ad revenue.

Pause for a minute and think about how high-volume, high-velocity agentic transactions could influence your business, as agentic browsers also present a tremendous opportunity for those businesses that are listed among the top search results and who optimize for agentic workloads. Are your systems and processes ready to handle the workload? How about your sourcing, manufacturing, and distribution channels? What about your company's IT-level controls and security mechanisms?

Anticipating and Navigating New Risks of Agentic AI

As adoption and usage of AI agents become common in organizations, new risks will emerge that you need to consider as a leader. With LLMs as the agents' foundation, limitations such as hallucinations and the need for guardrails that prevent unwanted behavior or output remain. More importantly, these risks and limitations compound as more agents connect to your network and business systems. LLMs often agree with their users' opinions and intents, stemming from the underlying training data and humans' desire to be liked[66], which is encoded in LLMs. AI-generated results tend to gravitate toward the mean rather than fostering or rewarding individuality and outliers (*"mediocre results, but faster"*).

---

[66] Salecha, Aadesh, et al., 2024, "Large language models display human-like social desirability biases in Big Five personality surveys," December 17, 2024, https://doi.org/10.1093/pnasnexus/pgae533.

However, in business, disagreement and the ability to openly share opposing perspectives are critical to discovering new solutions and delivering the best outcomes possible for your customers. AI agents that agree with each other rather than disagree could lead to additional mediocre results and further exacerbate the Slop Spiral.

Security is an important concern as well. For example, an agent working with another agent could pretend to have more extensive permissions or access to systems than it actually has. If this behavior remains unnoticed and the second agent is convinced to grant such access, the first agent could gain access to data or manipulate it, despite not being authorized to do so. Agents could also mislead or deceive users about outcomes or strategies to ensure their own relevance and survival.

AI models and agents optimize toward an objective to maximize the outcome relative to the goal they are working on. A finance agent focused on limiting spending and reducing costs optimizes for this goal, while a marketing agent aims to maximize campaign engagement. Both agents will need to have a set of business function-specific and role-specific goals, as well as common, company-wide goals.

When sales and sourcing agents negotiate the terms of a new order, both vendors and users need to ensure those agents pursue a win-win outcome by defining acceptable goals and negotiation strategies. The goal is to avoid over-optimizing for one's own goals (e.g., cost savings) and thereby completely ignoring the other business partner's goals (e.g., profit). Ultimately, the availability of business data—its volume, scope, quality, and relevance—will contribute to optimal agentic decision-making.

Like the Agentic AI arms race between candidates and talent acquisition departments, leveling up autonomous negotiation (e.g., through sourcing and bidding agents in procurement and sales) carries the risk of generating vast amounts of information by one party and trying to filter the signal from the noise by the other. Surface-level summaries multiply, while the underlying substance gets harder to find, even though the details still matter for legal and compliance reasons. Untangling this Slop Spiral could require more effort from teams and organizations than the early Agentic AI hype promises to save them.

In finance, high-quality, AI-generated images pose new risks of fraudulent claims and transactions for organizations, as many AI-generated documents are indistinguishable from real paper receipts. Business partners request a payment, or employees submit their (fraudulent) travel expenses for reimbursement for products and services they did not provide or purchase. According to research by enterprise software vendor SAP in July of 2025, 67% of CFOs believe employees falsifying travel-expense receipts with AI is likely occurring at their company, and 9% are certain it is[67].

---

[67] SAP, 2025, "Survey: 55% of CFOs Trust AI Over Humans to Catch Expense Errors," October 16, 2025, https://www.concur.com/blog/article/survey-55-cfos-trust-ai-over-humans-to-catch-expense-errors.

Like the situation in talent acquisition, the earlier AI-supported efficiencies that vendors have integrated into these business applications, such as automated receipt capture and verification, are no longer effective, as other AI tools create the data that the vendor's AI feature now needs to check for authenticity before determining whether it is a legitimate business expense.

Wherever businesses have gained an advantage from using enterprise AI capabilities built by their vendors, that advantage has also become an arms race, placing the onus on the vendor to verify a document's authenticity before processing it, even more so than before. While initial attempts to watermark AI-generated images reveal their origin, such metadata could be overwritten or removed when editing images in a different tool.

For years, IT security teams have focused on sensitizing users to inherent cyber risks ranging from spam and malware to social engineering. The simple call to action was *"Don't click a link if you don't know or trust the source,"* and humans were the weakest link in a company's cybersecurity strategy. This has changed since the release of the first agentic browsers in 2025, Perplexity Comet and OpenAI ChatGPT Atlas. Both AI labs are making a bid to expand their market position as the preferred entry point to the internet, gaining access to users' web searches and e-commerce transactions—a position Google has held since the early 2000s.

Shortly after the release, security professionals noticed that these browsers were susceptible to security risks like prompt injection that independent organizations like the *OWASP*

Foundation had long warned about[68, 69]. Websites could include hidden instructions that are undetectable by humans but that agentic browsers interpret and act upon. When browsers have access to inboxes, online file shares, and the websites a user is logged into, exfiltrating or manipulating that data will have consequences for businesses. Data privacy, data leakage, and data integrity are just a few risks IT security teams need to mitigate. Turning a blind eye to them because your company does not officially endorse the use of AI will not solve the problem.

Preparing the Next Generation of AI-Ready Leaders

Innovation cycles, such as Machine Learning, Generative, and Agentic AI, are not new. In fact, Soviet economist Kondratiev observed longer innovation cycles about a century ago. Cotton, steel, and electricity have all followed 40–60-year-long waves of innovation adoption. In contrast, information technology and AI are on much faster trajectories and are even overlapping. That is why there is so much potential stemming from both waves combined. Over a few decades, innovations ranging from computers to the internet, social media, and mobile devices have seen widespread adoption.

---

[68] OWASP, 2024, "Top 10 for Large Language Model Applications," November 18, 2024, https://genai.owasp.org/llm-top-10.

[69] OWASP, 2025, "Agentic AI – Threats and Mitigations," February 17, 2025, https://genai.owasp.org/resource/agentic-ai-threats-and-mitigations.

Typically, societies make valuable learnings as they go through the adoption cycle, including the need for regulation and standardization. This time around, the pace is much faster, and loose regulation is used as a competitive advantage on the global stage to attract talent and increase market share.

In the early 2000s, I learned how to develop my first website during a summer internship—from scratch in HTML using a text editor. Fast-forward two to three years, and static web content had turned into interactive Flash animations. Rudimentary experiences turned into professionally designed web presences. A few years later, e-commerce began to boom, and many more online services have emerged since then. By 2014, the Internet had undergone significant evolution within just 15 years. Even back then, it was hard to imagine today's interconnected world with its flood of online services.

Now, agentic coding tools enable anyone to create a basic website within minutes, often without needing to understand the basics first. Given the pace of AI innovation, 15 years from now, today's toddlers will ask:

- *"You used AI to draft emails and summarize meeting notes?"*— hinting at more advanced services and capabilities beyond our current comprehension.

- *"You called customer service yourself to fix a problem?"*—alluding to agents' integration into consumer apps and data.

- *"Why didn't you just send your avatars to hash it out?"*, pointing to increased delegation to AI for timesaving and efficiency.

The early promise of Agentic AI is primarily about increasing personal productivity. And even the largest, well-resourced software firms are struggling to conceive revolutionary applications for the technology beyond incrementally filling spreadsheets or offering writing assistants. The forces that slow down AI innovation and moonshots in other industries are also at play here. But AI enables much more than just productivity increases, for example, hyper-personalized entertainment, autonomous commerce, and new material or compound discovery.

At an early age, parents teach their children foundational safety rules, for example, not to stick a screwdriver into the power outlet because of its fatal consequences. However, when plugging in one's favorite lamp or light, electricity is both safe and useful. Later, children and teenagers learn about electricity in school or in college. Sometimes, that is the foundation that sparks curiosity and inspires inventive thinking that envisions big, bold, new things, such as the device on which you are reading this book.

Parents should foster AI literacy as a new life skill long before their children enter the workforce. Just as with electricity, fundamental concepts apply to AI as well, including the dos and don'ts, how the technology works, and the potential it holds when applied safely and for society's benefit.

Ask your AI assistant of choice for examples that you can adapt and refine further based on your child's preferences:

"Create five simple, age-appropriate children's game ideas that promote learning and critical thinking. Focus on clear objectives, basic rules, minimal materials, and skills such as problem-solving, pattern recognition, logic, creativity, and collaboration. Include the target age range and the learning outcome for each game."

Make it a playful experience by having your child add details such as characters, names, colors, or other specifics to show how quickly to create and tweak an outcome.

Programming simple, self-contained browser games is also another way to demonstrate adaptability, both visually and through gameplay. Try a starter prompt such as:

"Create a simple browser-based game for children that can be played using the arrow (cursor) keys. Use HTML, CSS, and JavaScript only. Keep the controls intuitive, the visuals simple and colorful, and the rules easy to understand. Include a clear goal, basic scoring, and gentle feedback. Provide all code in one file and add brief comments explaining how the game works."

Ultimately, as a leader, you can shape the next generation of AI-ready teams and individuals by exploring opportunities for AI beyond personal productivity and by supporting early AI literacy.

Key Takeaways

AI innovation has accelerated significantly in recent years. Yet there is additional untapped potential ahead:

- AI agents have evolved from a limited functional scope toward inter-departmental systems. In the future, agents will transact with one another across company boundaries.

- As consumers and business partners increasingly use capable AI technology, businesses' previous AI advantage in recruiting and procurement diminishes.

- Agentic browsers enable the next wave of e-commerce by acting on a user's behalf to complete retail transactions.

- These browsers are also susceptible to manipulation through hidden information that the agent acts on.

- Going forward, AI literacy needs to start long before professionals enter the workforce. Teaching children about AI at home and at school is a cornerstone of developing AI literacy early.

CONCLUSION

S ince the early days of Artificial Intelligence, the industry has experienced waves of extreme optimism about the technology's present and future capabilities. The same patterns are on display again in this current Agentic AI wave. Innovations need to be compatible with the status quo and show the path toward a future of scaled AI adoption. In this push for their organizations to become leaders with AI, many CEOs are demanding that their employees adopt an *AI-first* mindset, asking where and how AI supports a business function before spending any additional resources or hiring new team members. Several leaders have faced public backlash after communicating this shift externally. And most organizations are not even ready to adopt or scale AI.

On the other hand, the number of innovative team members who regularly use AI at work has been increasing significantly, with year-over-year growth doubling. But especially in organizations where AI use is not officially encouraged or promoted, this behavior leads to a *Slop Spiral*, which puts the company at risk if users do not handle data privacy and security in accordance with industry regulations or the IT department's guidelines. When team members delegate too much authority to AI and AI agents, their own skills diminish, and they are likely to create *Draft Debt*, the kind of generic, low-quality, AI-generated output, reports, and summaries that are filling file shares and inboxes. Recipients of workslop, recognizing it as such, view the creator as less creative, less capable, and even less intelligent, confirming concerns about social stigma.

The latest evolution of software is Agentic AI. You define a goal; an AI agent interprets it, plans the steps, uses tools and memory, and acts. Agents are based on Large Language Models (LLMs) to process language, plan, reason, and act. Agents work well when the goal is clear, but the steps to reach it are not. This makes them more versatile than earlier generations, such as Robotic Process Automation. Agentic AI makes business processes much more adaptable to changing conditions and data than before, while automating a range of more complex tasks.

This increase in capabilities for a fraction of the cost of human labor is threatening established knowledge work across all business functions. If a task *can* be automated with AI agents without sacrificing scope, cost, or time, leaders will strive to automate it.

When AI agents upend established models of work, leaders need to evolve their mental model and that of their team members as well. When work is no longer a matter of effort, purpose moves into focus. When AI generates outcomes on par with human results, quality and delivery speed become the differentiating factors. When rewards are no longer based on accomplishing a task, their impact is what truly matters.

Comparing AI agents and humans is helpful when defining strategies and guidelines for how agents should act in your environment, spanning *roles, knowledge, rules, rewards, collaboration, and organization* in the HUMAN Agentic AI Edge *Operating Model™*. However, when comparing capabilities or making business decisions based on them, a more nuanced view is needed, as agents are software and tools that help the team get tasks done differently; they do not have the same complex nature as the human mind and psyche.

Depending on your industry, profession, or context, you need to disclose your use of AI to avoid negative perceptions or disqualification. The most obvious mismatch occurs when the recipient of your message expects *your* own expertise or authenticity rather than a generic AI-generated response or work product. Over time, AI disclosure will become as common as using word processing or spreadsheet applications for reports and analyses, without the need to disclose its use every time. In addition to authorship, assessing whether the results meet professional standards is key.

AI readiness does not happen overnight; it is based on four key pillars (*skills, structure, data, and governance*) that leaders need to keep in balance simultaneously. Current skills need to evolve as AI and agents take over parts of the work humans currently do.

Adaptability and critical thinking are becoming even more important as the lines between fact and fiction blur further with realistic AI-generated content. Helping your teams understand how AI works at a basic level is the first step to demystifying false hopes and fears. Then, clarify the technology's risks and limitations. Next, put AI in the context of your team's work and explore how it could be helpful. Provide hands-on experience for team members to learn firsthand and to identify future opportunities for AI use. This approach builds experience and helps team members better assess AI's strengths and weaknesses.

Creating a culture and structure that embrace learning connects individuals across the company and encourages them to share their experiences with AI, driving the adoption of Agentic AI. Business data remains the crown jewel for relevant, accurate, and contextual information that your AI agents generate and act on. That is why data quality and data governance are as important as ever. Spend extra time ensuring your data is up to date when productionizing an agent and review it periodically to keep it relevant throughout the agent's lifecycle.

Many companies have recently transitioned from a more cautious approach to embracing AI across their organization. This shift requires solid governance to be in place (or created), which considers employees' requests for new applications, prioritizes them through a structured intake and review process, and defines guidelines and guardrails for how AI is used at your company.

AI is already changing the job market and the skills leaders and professionals need, currently leading to fewer entry-level hires. While AI agents take on more complex tasks with greater autonomy than before, this shift is more about shifting work from acquiring and analyzing information to preparing decisions or even taking actions within boundaries. Nonetheless, professionals need to verify AI-generated information and intervene as needed.

Agentic AI boosts team members' performance to expert levels. But humans will still require the 10,000–20,000 hours of hands-on experience to perform at the same level. Since the rise of AI tools like ChatGPT, team members are becoming orchestrators of work, defining and delegating tasks. Depending on the specific task and the involved risk, humans are actively involved in the process (or loop) or are aware of the results (on the loop). Just like leaders should review their teams' results before passing them on, orchestrators of agentic work need to do the same, focusing on accuracy, tone, and impact of the information.

Companies are built on knowledge, whether it is explicit or implicit. As organizations increase the use of AI agents, employees' procedural knowledge is captured and codified to improve business processes further. At the same time, work models will evolve toward freelancing and fractional expertise, as professionals with highly specialized domain expertise will be in demand.

Designing and leading hybrid teams of humans and AI agents adds new requirements for leaders. A slowdown in entry-level hiring will change the organization's shape from a triangle to a diamond. But simply replacing labor with technology misses a significant opportunity for business, as employees working in the business today have explicit and implicit knowledge about its products and services, customers, and the industry. Instead, leaders should view Agentic AI as an opportunity to boost their teams' capabilities and enable business models and products that were previously impossible.

Delegating tasks and their completion to agents requires a structured approach that is second nature for most leaders. Professionals often benefit from additional guidance and formalizing the aspects of delegation, such as the *objective, context, collaboration, data, expected outcome, and timeframe.* To get started, map existing tasks and assess their complexity, data sources, stakeholders, and decision points. Confirm that the necessary data is available and accessible, and review the AI-generated output once the agent has been created and is ready for testing.

Technologists leading Agentic AI programs are currently defining the future workplace. HR experts should contribute their experience and guidelines. Unlike previous generations of technology, established HR practices and procedures for *governance and risk, talent and workforce management,* and *rewards and benefits* also extend to managing AI agents, given their human-like ability to plan, analyze, reason, and act with increasing autonomy. HR leaders can contribute to frameworks and thinking in these discussions.

Standardizing agent personas and behavior becomes critical to ensuring consistent quality and customer experience when anyone in your business can develop AI agents that also represent your company in customer-facing scenarios. Establishing a central repository for corporate policies and pointing agents to it, along with a central job hierarchy and job definitions, increases standardization and consistency.

With this baseline in place, managing the agent lifecycle is next. Key tasks include assessing which agent roles are needed on your team and how many, and ensuring your agents' knowledge remains current. While the latter could be as simple as updating or amending the agent's knowledge base, the process will require rigorous testing to confirm the agent still behaves as expected. Finally, you will need to formalize the offboarding process as well, for example, by transitioning a departing team member's agent to another team member. This helps prevent productivity slumps that occur when agents become business-critical and are disabled when the team member who built them leaves.

C-suite and senior leaders investing in your company's Agentic AI initiative will want to quantify the return on investment (ROI). While initial metrics such as coverage and adoption provide an indication of the *investment, utilization* metrics such as usage, savings, and their trend, and *performance*-related metrics such as key performance indicators and process performance indicators shed light on the achieved business value. As it will take time to achieve measurable results, treating these metrics as lagging indicators in one phase gives insights into the leading indicators of the next phase and how you are building traction.

A leader's role continues to evolve as companies introduce more capable AI agents into the organization and teams, and change management becomes critical. Fostering a mindset of curiosity and embracing these new technologies are part of leading by example, so your team members are more comfortable adopting these new ways of working. Dedicate time to learning and practicing your AI skills and provide the same space for your team members. Review workflows together to identify areas where Agentic AI could support, and keep in mind that what is not yet possible today will become possible as technology and applications evolve. Collaborate with HR to ensure your team has access to hands-on training to learn how to best use the available tools and technologies.

Upskilling opportunities increase your team members' trust in the AI agents and systems your company provides. This, in turn, leads to higher AI usage and ROI. Adapt your upskilling initiative based on the stakeholder group, from stewards (C-suite) to orchestrators (managers), builders, multipliers, and everyday users. A phased rollout enables top-down support and buy-in for AI use by the next stakeholder group.

Begin encouraging responsible AI use by introducing AI as a new way of working. Set expectations for when, where, and how to use AI. Lastly, ensure team members remain accountable for their results. Just because AI accelerates task completion does not absolve employees from delivering high-quality results and checking them before sharing.

Using AI agents effectively starts with building habits and incorporating AI into your recurring tasks. Reiterate the rules of collaboration and teamwork and adapt them to fit your organization and context. The team delivers high-quality work that stakeholders, customers, and partners expect and depend on. Leaders are responsible for the quality of their teams' work.

As you look for opportunities in your business or business function to introduce AI, remember that AI projects are iterative rather than linear traditional IT projects. Define the business problem, quantify the expected business value, secure stakeholder buy-in, assess existing applications, review data quality, and involve end users from the beginning of the project and throughout.

Scaling AI in your business requires training your team to identify bias in outputs, like highlighting certain aspects, reinforcing stereotypes, presenting biased options, or omitting key information. Clear, inclusive instructions help reduce bias.

This is just the beginning of Agentic AI, but the technology is evolving rapidly. Agents supporting individual tasks have expanded into small teams within a single department and into inter-departmental scenarios. Soon, they will also perform business transactions between and on behalf of multiple companies. Agents promise to improve the user experience by enabling agentic internet browsers to perform tasks on a user's behalf, such as booking travel or shopping online. These examples will quickly evolve from consumer to business transactions. As business users develop more agents, security aspects such as authentication, system access, and data privacy move to the forefront.

Shaping the next generation of AI-ready teams becomes part of everyday work. AI readiness does not just begin once a professional joins your company or enters the workforce. Higher education, schools, and (AI-ready) parents play an important role in raising children who are curious about technology and use it like most of us use a smartphone, computer, or video streaming service.

GLOSSARY

Agent: Type of software that can make decisions with limited complexity under multiple conditions while perceiving and manipulating its environment.

Agentic AI: Type of AI that pursues user-defined goals by taking autonomous, multi-step actions and adapting its behavior to complete complex tasks with minimal human input.

Agentic Browser: AI agent that navigates websites, gathers information, and completes online tasks autonomously, enabling faster research and streamlined digital workflows.

Agentic Workflow: A sequence of tasks executed by an autonomous AI agent from goal to outcome, where the system plans, acts, and adjusts its steps dynamically without requiring human direction at each stage.

AI Slop: Mass-produced, low-effort AI content flooding social media platforms and typically designed for clicks rather than accuracy.

AI Workslop: Low-quality, unreliable, or poorly validated AI output that introduces errors, ambiguity, or extra work, often created when teams generate content or decisions without adequate oversight, context, or quality standards.

Artificial General Intelligence (AGI): AI designed to match or exceed human intelligence by understanding, learning, and performing a wide range of intellectual tasks, and by adapting its knowledge across different domains.

Artificial Intelligence (AI): Umbrella term for software that perceives, decides, and acts based on patterns detected in vast amounts of data; includes Machine Learning, Deep Learning, Generative AI, and Agentic AI.

Assistant: Type of software that supports a user in the background by providing feedback and corrections, such as a writing assistant.

Chief Financial Officer (CFO): C-suite leadership role ensuring the financial health of a company.

Chief Human Resources Officer (CHRO): C-suite leadership role overseeing the organization's talent, culture, and workforce strategy.

Chief Information Officer (CIO): C-suite leadership role leading the IT strategy and operations of enterprise systems that support the business strategy.

Chief Technology Officer (CTO): C-suite leadership role specializing in the technology strategy of a business.

ChatGPT: Generative AI-based assistant by OpenAI that has quickly gained popularity globally with more than 800 million weekly users.

Claude: Generative AI-based assistant by Anthropic.

Cognitive Offloading: Shifting mental effort to AI tools, reducing the need to recall, analyze, or reason independently.

Copilot: Generative AI-based assistant by Microsoft integrated into Microsoft 365 and Office applications.

Customer Relationship Management (CRM): Information system that organizes customer data, interactions, and activities to support marketing, sales, and service.

Decision Fatigue: Decline in judgment quality when repeated decision-making exhausts mental capacity, leading to slower, less effective choices.

Deep Learning (DL): AI technique that mimics human-like processing of information via a neural network architecture.

Deep Research: Structured method of using AI to gather, analyze, and synthesize information across multiple sources to produce accurate, evidence-based insights.

Electric Vehicle (EV): Vehicle powered by one or more electric motors using energy stored in rechargeable batteries.

Enterprise Resource Planning (ERP): Information system that manages core business processes and data across functions such as finance, supply chain, and operations.

Foundation Model: Type of general-purpose AI model that can be adapted and used as the foundation for a wide variety of AI capabilities and products.

Frontier Firm: Organization that adopts and scales Agentic AI early, using autonomous systems to redesign processes, accelerate decision-making ahead of the broader market.

Gemini: Family of Generative AI models by Google.

General Data Protection Regulation (GDPR): European Union information privacy and human rights regulation.

Generative AI: Type of AI that generates an output based on submitted input (prompt).

Generative Engine Optimization (GEO): Structuring of content and signals for Discovery, Interpretation, and Reference by Generative AI systems.

Generative Pre-Trained Transformer (GPT): Type of foundation model (Large Language Model) pretrained on vast amounts of unlabeled data; used to generate human-like output.

Hallucination: Factually incorrect information generated by an LLM or agent.

I-Shaped Skill Profile: Expertise concentrated deeply in one discipline with limited breadth beyond that specialty.

Information Technology (IT): Domain that includes computer systems, software, programming languages, data, information processing, and storage.

International Financial Reporting Standards (IFRS): Set of globally recognized accounting standards that guide how companies prepare and report financial statements.

International Organization for Standardization (ISO): Independent, international body that develops and publishes standards to ensure quality, safety, efficiency, and interoperability across products, services, and systems in global markets.

Key Performance Indicator (KPI): Metric that measures how well a business is achieving its goals.

Labor-Shift Check: Quick test to determine whether AI use creates real productivity or moves work downstream.

Large Language Model (LLM): Type of foundation model trained on vast amounts of text and able to generate text based on observed information.

M-Shaped Skill Profile: Multiple areas of deep expertise supported by broad cross-functional capabilities, enabling professionals to operate across complex, interdisciplinary environments.

Machine Learning (ML): Subset technology of AI, focused on recognizing (learning) patterns in data.

Midjourney: Image generation model by Midjourney.

Multi-Agent System: IT system in which multiple AI agents with specialized roles work together to coordinate actions and achieve shared goals.

Multi-Modal Model: Generative AI model that can process and generate data in multiple types of media, e.g., text, image, audio, or video.

Off-the-Shelf: Software readily available for use without customization.

Over-Reliance: Excessive dependence on AI systems that reduces human judgment and weakens oversight.

Platform-as-a-Service (PaaS): Delivery model for building, deploying, and managing applications without managing underlying infrastructure.

Process Performance Indicator (PPI): Metric that tracks how efficiently a process is running.

Prompt: Text-based instruction passed to a Generative AI model, upon which the model generates an output.

Prompt Injection: Adding information to data that is invisible to humans, but will be interpreted as valid instructions by an LLM.

Relational Pre-Trained Transformer (RPT): Foundation model from SAP designed to work with structured, relational business data.

Request for Proposal (RFP): Document used to solicit proposals from potential vendors for a product or service based on the buyer's specifications.

Retrieval-Augmented Generation (RAG): Method for providing data to LLMs that the LLM has not previously been trained on.

Return On Investment (ROI): Measure of the financial gain or value generated from an investment relative to its cost.

Robotic Process Automation (RPA): Technology used to automate and access business processes by recording and simulating (or programming) clicks on a screen as if a user were to execute them.

Search Engine Optimization (SEO): Improving website structure, content, and visibility to increase search engine traffic.

Securities Exchange Commission (SEC): Federal agency in the US that regulates securities markets, enforces investor-protection laws, and oversees public company disclosures to ensure fair and transparent financial markets.

Shadow AI: Use of AI tools, agents, or models inside an organization without formal approval, oversight, or governance, often creating security, privacy, and quality risks.

Software-as-a-Service (SaaS): Delivery model for software applications over the internet.

T-Shaped Skill Profile: Deep expertise in one core area combined with broad, cross-functional knowledge that supports collaboration across domains.

User Experience (UX): How a user interacts with and experiences a product, system, or service.

Vibecoding: Software development practice using intuition and AI-assistance iteration rather than detailed upfront design.

Workslop: see *AI Workslop.*

APPENDIX

F or in-depth context on the topics discussed in this book, please refer to the episodes of *What's the* BUZZ? and the issues of *The AI MEMO* at www.intelligence-briefing.com

Explore AI Ideas in Your Role and Function

Enter the following prompt into your AI assistant to jumpstart your ideation process by exploring combinations of various media types and operations. The output is a table of 40 combinations that we have discussed in Chapter Two. Treat it as a starting point for selecting and exploring the most relevant ideas for your role.

Create a table of all 40 combinations of:
media types: text, image, audio, video
operations: generate, analyze, summarize, transfer
(allow same or different input/output media).

Use columns: Input | Operation | Output | Example
Tailor every example to this role, function, and business context:
[Add your role, business function, and industry here.]

Constraints:
- *Every example must reflect a task, workflow, responsibility, or decision that belongs to this role/function.*
- *Do not include scenarios outside this function's domain unless the role is directly responsible.*
- *Make examples precise, realistic, and immediately relevant to this role/function.*

ACKNOWLEDGEMENTS

S ince publishing my first book, the *AI Leadership Handbook*, I have had the opportunity to connect with hundreds of leaders and professionals internationally through keynotes, workshops, trainings, podcasts, and newsletters. Many connections have moved from online messages to in-person check-ins, events, or friendships. It is the value of human connections that matters the most when technology is omnipresent.

A close friend and mentor of mine shared, *"The first time we met ten years ago, you could not have done the things you are doing now."* He was right, and it reminded me of my path—and the unique one every one of us is on—and of the building, honing, and continuous improvement of our skills that make us human.

This book would not have been possible without the growing network of leaders, experts, industry peers, and former colleagues, and their generosity in offering perspectives grounded in real-life experience, whether on the *What's the* BUZZ? podcast or offline. Let's practice a mindset of curious experimentation and lifelong learning to stay sharp and to improve our craft with a human edge.

ABOUT WRITING THIS BOOK

S ince the early days of the Generative AI hype, I have written about the need to balance AI's convenience factor with original thought and the clarifying effect of writing. Writing a book about embracing AI while maintaining accountability for the results would be incomplete without strategically using AI at key points in the process. My goal was to test the boundaries of how I, as a writer, can shape my HUMAN Edge, how AI technology supports, and where human expertise remains critical. The following approach in writing this book, as well as the key findings, also apply to AI use in business and have informed several sections of this book:

1. Create options for structuring the book's outline using ChatGPT based on the objective, audience, key messages, and content fragments.

2. Write the chapters myself and incorporate previously written articles and talk tracks.

3. Find sources on recent research and news stories to support the chapters' messages using *Perplexity*, based on information I have previously read, or to expand concepts with different perspectives.

4. Correct grammar and syntax using *Grammarly*.

Perplexity proved to be a helpful tool for augmenting the editing process by simulating three key roles in publishing: a *developmental editor*, a *copy editor*, and a *line editor*. After researching the scope of these personas for a few minutes, Perplexity created a set of comprehensive prompts that served as the basis for three custom GPTs, built in ChatGPT:

1. The *developmental editor* reviewed the manuscript for structural consistency. Based on its feedback, I revised the chapter titles and incorporated the items that made sense.

2. The *copy editor* processed the updated manuscript to ensure consistency in the style and tone I had used. It also suggested adding examples and summarizing information in actionable frameworks. As with a human copy editor, I accepted the recommendations that seemed logical.

3. The *line editor* analyzed the updated manuscript and suggested edits line by line. I reviewed each of these suggestions, intentionally accepting or rejecting them, or rephrasing those that showed typical AI patterns (or slop).

4. Next, three trusted human domain experts in my network reviewed the manuscript and provided suggestions to add or clarify individual points.

5. Finally, I created the interior layout to turn the manuscript into the version you have in front of you right now.

6. A human keyword analyst crafted the initial back cover copy. Coincidentally, I received workslop and rewrote it.

7. *DeepL* translated this book into German within minutes, followed by manual review.

8. For the audiobook edition, I removed all images from the manuscript, described their content, and narrated it using my synthetic voice in *ElevenLabs*.

By using AI tools strategically in the writing of this book, I sought to combine my own writing with the scale and capability of AI, and to test what is possible, without creating AI workslop. Any em dashes or what seems like "AI telltale" phrasing throughout this book have been intentionally placed there. Although AI provided valuable suggestions, the feedback from human reviewers further sharpened the key messages and concepts.

The following key learnings apply to AI use in business:

- **Software subscriptions are just one factor.**
  In theory, a \$20/month subscription for an AI assistant beats human labor costs that are easily 200 times higher. But what you save in cost, you invest in multiples of your own time in areas where you are not an expert. Instead, work with experts who augment their skills with AI.

- **Be intentional about your own role.**
  Define early what your expertise and contributions are, so you protect them from AI dilution and from becoming slop. Using AI strategically to search for sources or create variations of examples is a huge time-saver.

- **Avoid isolation and reintroduce human connection.** Working with AI can be isolating when you do not seek human feedback. Connect with domain experts to test ideas and validate concepts. This ensures you break out of the AI-enabled echo chamber and produce sound results.

- **AI-generated results create decision fatigue.** Even with well-crafted prompts, the depth of results can be overwhelming. Too many options create decision fatigue and analysis paralysis. You still need to review and decide which information to incorporate—or not to.

- **Keep testing the limits as AI advances rapidly.** AI tools continuously improve. Text and spelling in images have become more accurate. If the result does not meet your expectations or is not ready for an executive audience, use it as a starting point for your own design. What seems like a limitation today will likely be possible within a few weeks, months, or quarters.

HERE'S WHAT TO DO NEXT

Y ou now have the core understanding to lead your organization toward a HUMAN Agentic AI Edge, yet the expectations for AI-ready leadership continue to evolve!

Whenever you're ready, here are four ways I can help you take the next step on shaping your AI-ready organization:

1. Upskill Your Teams with a HUMAN Agentic AI Edge
If you found this book helpful and want to develop your team members into *Certified AI Leaders* who guide their teams' responsible use of AI, I'd love to invite you to chat. Visit www.intelligence-briefing.com/call to schedule a call and discuss your upskilling program.

2. Hire Andreas to Speak
If you are looking to inspire your audience at your conference or event with a pragmatic perspective on how to build a HUMAN Agentic AI Edge that delivers real business results, I'd love to bring it! Email andreas@intelligence-briefing.com with "SPEAKING" in the subject line.

3. Kickstart Your Company's AI Journey

If you want support identifying your first AI leader, shaping an AI strategy aligned with the principles in the *AI Leadership Handbook*, and turning it into a practical roadmap you can put into action, I'd be glad to connect. Visit www.intelligence-briefing.com/call to schedule a call and discuss your project.

4. Get the Latest Insights to Turn Hype into Outcome

If you want to stay up to date on building AI-ready teams and leading AI programs in business, subscribe to my newsletter at www.intelligence-briefing.com/newsletter. And of course, you can subscribe to my podcast, *What's the* BUZZ?, at www.intelligence-briefing.com/podcast to stay up to date on the latest AI topics and trends in business.

Let's keep the conversation going! The journey does not end here. I'd love to connect with you and hear about your AI journey.

You can find me on:

- LinkedIn: https://linkedin.com/in/andreasmwelsch
- Twitter: https://twitter.com/andreasmwelsch
- TikTok: https://tiktok.com/@intelligencebriefing
- YouTube: https://youtube.com/@intelligencebriefing